Simulation Foundations, Methods and Applications

More information about this series at http://www.springer.com/series/10128

Khalid Al-Begain · Andrzej Bargiela
Editors

Seminal Contributions to Modelling and Simulation

30 Years of the European Council of Modelling and Simulation

Editors
Khalid Al-Begain
Faculty of Computing, Engineering
and Science
University of South Wales
Pontypridd
UK

and

Kuwait College of Science and Technology
Kuwait City
Kuwait

Andrzej Bargiela
ECMS
Nottingham
UK

ISSN 2195-2817 ISSN 2195-2825 (electronic)
Simulation Foundations, Methods and Applications
ISBN 978-3-319-81602-9 ISBN 978-3-319-33786-9 (eBook)
DOI 10.1007/978-3-319-33786-9

Printed on acid-free paper

This Springer imprint is published by Springer Nature
The registered company is Springer International Publishing AG Switzerland

Preface

It is a great privilege for anyone to be part of any event with historic significance. This year the European Conference on Modelling and Simulation reaches the 30th milestone and I have the honour of being the President of the European Council for Modelling and Simulation as the scientific organization behind the conference. To mark this significant occasion, Springer Verlag is publishing this memorial book of selected high-impact articles from the keynotes and best papers of the last 5 years.

Modelling and simulation have become an integral part of every design process from the planning stages all through to the evaluation and validation of implementation and even beyond. Simulation has been the methodology used to designing vehicles for space explorations, in testing ideas for industrial processes and products and to understand human and living objects behaviour. It is not far from truth to state that every researcher in science, engineering, economics or medical science must have some involvement with modelling and/or simulation at various levels.

Simulation in Europe has long history and traditions and the modelling and simulation community has played significant role in the advancement of the modelling and simulation science and methodologies and effective application of the methodologies in numerous fields. The Modelling and Simulation community in Europe and as part of the international research community expressed its presence in various forums and conferences. One of the most important forums is the European Conference on Modelling and Simulation (ECMS) which is the international conference dedicated to help define the state of the art in the field. For the last 27 years, ECMS has proven to be an outstanding forum for researchers and practitioners from different fields involved in creating, defining and building innovative simulation systems, simulation and modelling tools and techniques, and novel applications for modelling and simulation. In the first chapter of this book, Professor Eugene Kerckhoffs, the first president of the European Council for Modelling and Simulation (known then as the Society of Computer Simulation—Europe) gives an account on the history of the council and the conference.

Over the last 29 episodes of the ECMS, generations of great scientists presented their work on modelling and simulation. Some were contributing to the advancement of the discipline itself, while other presenting significant contribution to other sciences and disciplines using simulation. In particular, the conference series witness countless significant keynote talks presenting significant and cutting-edge results in many areas that would require a multi-volume book to report on.

In this book, the editors opted to highlight a significant and exciting area in the application of modelling and simulation, namely the modelling of human brain. This is a common theme among the first three contributions in the book in Chaps. 2–4.

In the seminal talk in ECMS2014, Professor May-Britt Moser from the Norwegian University of Science and Technology and Nobel Prize Winner in 2014 in Physiology or Medicine, presented the results of her research on how the brain controls spatial navigation in mammals by activating functionally specialized cell types in the medial temporal lobe. A brief account of the content of the talk is given in Chap. 2.

Related to the topic, Professor Steve Grossberg reported on further research on medial entorhinal grid cells and hippocampal place cells are crucial elements in brain systems for spatial navigation and episodic memory. In his extended Abstract in Chap. 3, he summarizes the development of the GridPlaceMap neural model that explains many data about these cells and behaviours in a unified way. He also gives a thorough reference list to summarize and simulate these data. The chapter provides a high-level overview of the results, along with the unifying neural design principles and mechanisms that enable them to be realized.

Further account of significant advancement in the field is reported in Chap. 4 with the article by Professors Peter D. Neilson and Megan D. Neilson entitled “Modelling the Modeller”. The authors explain that the Adaptive Model Theory is a computational theory of the brain processes that control purposive coordinated human movement. It sets out a feedforward–feedback optimal control system that employs both forward and inverse adaptive models of (i) muscles and their reflex systems, (ii) biomechanical loads on muscles and (iii) the external world with which the body interacts. From a computational perspective, formation of these adaptive models presents a major challenge.

Moving out of the human body but still in relation with human behaviour, Professor Alexander H. Levis from George Mason University, USA presents in Chap. 5 a detailed account of his work on the modelling of a human organization for the analysis of its behaviour in response to external stimuli is a complex problem which requires development and interoperation of a set of several models. Each model developed using different modelling languages but the same data, offers unique insights and makes specific assumptions about the organization being modelled. Prof. Levis shows that interoperation of such models can produce a more robust modelling and simulation capability to support analysis and evaluation of the organizational behaviour.

As a significant contribution to the simulation methodology, Professor Martin Ihrig, University of Pennsylvania, presents new research architecture for the simulation era in Chap. 6. This chapter proposes novel research architecture for social scientists that want to employ simulation methods. The new framework gives an integrated view of a research process that involves simulation modelling. It highlights the importance of the theoretical foundation of a simulation model and shows how new theory-driven hypotheses can be derived that are empirically testable. The author describes the different aspects of the framework in detail and shows how it can help structure the research efforts of scholars interested in using simulations.

Further significant contribution to the modelling and simulation methodology is presented in Chap. 7 by Professor Witold Pedrycz from the University of Alberta with the title "Fuzzy Modeling and Fuzzy Collaborative Modeling: A Perspective of Granular Computing". The author elaborates on current developments in fuzzy modelling, especially fuzzy rule-based modelling, by positioning them in the general setting of Granular Computing. This gives rise to granular fuzzy modelling where the models built on a basis of fuzzy models are then conceptually augmented and made in rapport with experimental data. In the chapter, two main directions of granular fuzzy modelling dealing with distributed data and collaborative system modelling and transfer knowledge are formulated and the ensuing design strategies are outlined.

Staying with contributions to the methodology, Zuzana Kominkova Oplatkova and Roman Senkerik from Tomas Bata University in Zlin, Czech Republic present in Chap. 8 an article on control law and pseudo neural networks synthesized by the evolutionary symbolic regression technique. This research deals with the synthesis of final complex expressions by means of an evolutionary symbolic regression technique—analytic programming (AP)—for novel approach to classification and system control. In the first case, classification technique, a pseudo neural network is synthesized, i.e. relation between inputs and outputs are created. The inspiration came from classical artificial neural networks where such a relation between inputs and outputs is based on the mathematical transfer functions and optimized numerical weights. AP will synthesize a whole expression at once. The latter case, the AP, will create a chaotic controller that secures the stabilization of stable-state and high-periodic orbits—oscillations between several values of the discrete chaotic system. Both cases will produce a mathematical relation with several inputs; the latter case uses several historical values from the time series.

In January 2012, the world woke up on the news of a horrific accident in which a cruise liner called Costa Concordia hit a rock in the shores of Italy. This resulted is a significant loss of lives. As the investigations started, the Court was looking for methods to create better understanding of what caused the accident. It happened that the Italian group of researchers (Paolo Neri and Bruno Neri from University of Pisa and Paolo Gubian and Mario Piccinelli from University of Brescia) have developed a simple but reliable methodology for short-term prediction of a cruise ship behaviour during manoeuvres. The methodology is quite general and could be applied to any kind of ship, because it does not require the prior knowledge of any structural or mechanical parameter of the ship. It is based only on the results of manoeuvrability data contained in the Manoeuvring Booklet, which in turn is filled out after sea trials of the ship are performed before its delivery to the owner. The team developed this method to support the investigations around the Costa

Concordia shipwreck, which happened near the shores of Italy in January 2012. It was then validated against the data recorded in the "black box" of the ship, from which the authors have been able to extract an entire week of voyage data before the shipwreck. The aim was investigating the possibility of avoiding the impact by performing an evasive manoeuvre (as ordered by the Captain some seconds before the impact, but allegedly misunderstood by the helmsman). The preliminary validation step showed a good matching between simulated and real values (course and heading of the ship) for a time interval of a few minutes. Chapter 9 gives a full account of the work by the Research team as presented in ECMS conference.

In engineering it is usually necessary to design systems as cheap as possible whilst ensuring that certain constraints are satisfied. Computational optimization methods can help to find optimal designs automatically. However, the team of Lars Nolle (Jade University of Applied Science), Ralph Krause (Siemens AG) and Richard J. Cant (Nottingham Trent University) demonstrated in the work in Chap. 10 that an optimal design is often not robust against variations caused by the manufacturing process, which would result in unsatisfactory product quality. In order to avoid this, a meta-method is used in here, which can guide arbitrary optimization algorithms towards more robust solutions. This was demonstrated on a standard benchmark problem, the pressure vessel design problem, for which a robust design was found using the proposed method together with self-adaptive step-size search, an optimization algorithm with only one control parameter to tune. The drop-out rate of a simulated manufacturing process was reduced by 30 % whilst maintaining near minimal production costs, demonstrating the potential of the proposed method.

The last three chapters of the book comprise significant contributions to the analytical and stochastic modelling which were presented as part of the International Conference on Analytical and Stochastic Modelling and Applications that have been collocated with the ECMS conference.

In Chap. 11, Dieter Fiems, Stijn De Vuyst, and Herwig Bruneel, all from Ghent University discuss in Chap. 5 the packet loss problem which is an important and fundamental problem in queueing systems and their applications in modelling communications networks. Buffer overflow in intermediate network routers is the prime cause of packet loss in wired communication networks. Packet loss is usually quantified by the packet loss ratio, the fraction of packets that are lost in a buffer. While this measure captures part of the loss performance of the buffer, the authors show that it is insufficient to quantify the effect of loss on user-perceived quality of service for multimedia streaming applications.

Approximating various real-world observations with stochastic processes is an essential modelling step in several fields of applied sciences. In Chap. 12, Gabor Horvath and Miklos Telek from Budapest University of Technology and Economics, present fitting methods based on distance measures of marked Markov arrival processes. They focus on the family of Markov-modulated point processes, and propose some fitting methods. The core of these methods is the computation of the distance between elements of the model family. They first introduce a methodology for computing the squared distance between the density functions of two phase-type (PH) distributions. Later, they generalize this methodology for computing the distance between the joint density functions of k-successive inter-arrival times of Markovian arrival processes (MAPs) and marked Markovian arrival processes (MMAPs).

Chapter 13 of the book presents a very interesting modelling concept called "Markovian Agent Models". A Markovian Agent Model (MAM) is an agent-based spatiotemporal analytical formalism aimed to model a collection of interacting entities guided by stochastic behaviours. An MA is characterized by a finite number of states over which a transition kernel is defined. Transitions can either be local, or induced by the state of other agents in the system. Agents operate in a space that can be either continuous or composed by a discrete number of locations. MAs may belong to different classes and each class can be parameterized depending on the location in the geographical (or abstract) space. In this chapter, Andrea Bobbio, Davide Cerotti, Marco Gribaudo, Mauro Iacono and Daniele Manini (Italy) provide a very general analytical formulation of a MAM that encompasses many kinds of forms of physical dependencies among objects and many ways in which the spatial density may change in time.

As stated at the beginning of the Preface, the field of modelling and simulation cover huge range of theoretical and application areas that cannot be covered in one book. The articles selected in this volume represent a small sample of the contributions over 29 conference and they meant to mark the 30th ECMS conference. On this occasion it is important to mention few people whose contributions (scientific or organizational) allowed the conference and the council to reach this significant milestone. The list will definitely not be comprehensive but I would like to mention Professor Eugene Kerckhoffs, first President of the Council, Professor Andrzej Bargiela, Past President, Dr. Evtim Petychev, Past President and Mrs Martina-Maria Seidel, ECMS Officer Manager and the tireless driving force behind the organization of all ECMS conference. Specifically, for the help in editing this book, I would like to thank Professor Robin Bye from Norwegian University of Science and Technology and General Chair of ECMS2014 for his help in getting several contributions.

Finally, the European Council for Modelling and Simulation is very grateful to the Simon Rees, Associate Editor, Computer Science, in Springer Verlag for the initiative to publish this book and for the help provided all through the process.

Prof. Khalid Al-Begain
President of the European Council of Modelling and Simulation

Contents

About the Editor

Khalid Al-Begain is a Professor of Mobile Computing and Networking at the University of South Wales (Previously called University of Glamorgan) since 2003. He is currently the President of Kuwait College of Science and Technology, leading the establishment of a new highly ambitious private university in Kuwait.

Al-Begain is still the Director of the Integrated Communications Research Centre (ICRC) at the University of South Wales, UK. He has been the Director of the Centre of Excellence in Mobile Applications and Services (CEMAS), a £5 million Centre partly funded by European Regional Development Fund (ERDF).

He received his M.Sc. and Ph.D. in Communications Engineering in 1986 and 1989, respectively, from Budapest University of Technology, Hungary. He also received Postgraduate Diploma in Management from University of Glamorgan in 2011 after finishing 2-year MBA course modules.

He has been working in different universities and research centres in Jordan, Hungary, Germany and the UK. He has led and is leading several projects in mobile computing, wireless networking, analytical and numerical modelling and performance evaluation.

Al-Begain is the President of the European Council for Modelling and Simulation (2006–2010 and 2014–date). He also was the President of the Federation of European Simulation Societies (EuroSim) leading the UK Presidency term of 2010–2013. He is the first person ever to take this role in both the organizations.

Al-Begain was a member of the Welsh Government Ministerial Advisory Steering Group for ICT in Schools. He is an UNESCO Expert in networking, British Computer Society Fellow and Chartered IT Professional, Senior Member of the IEEE and IEEE Communications and Computer Societies, member of the UK EPSRC College (since 2006) and the expert panel member of Belgium Research Funding Council (FWO), Life Fellow of the German Alexander von Humboldt Foundation, Wales Representative to the IEEE UK&RI Computer Chapter management committee and has been the UK representative to the COST290 Action management committee (2005–2009).

Al-Begain has been also appointed as Adjunct Professor at the Department of Computing and Engineering at Edith Cowen University, Australia, and as a member of the Governing Board of the West Australia Centre of Excellence on Microphotonic Systems since 2005.

So far, Al-Begain has been the General Chair of 25 international conferences. He has been the General Chair of the International Conference of Analytical and Stochastic Modelling Techniques and Applications (ASMTA) since 2003 and, in particular, he is the Founder and General Chair of the IEEE International Conference and Exhibition on Next Generation Mobile Applications, Services and Technologies (NGMAST).

He is the Winner of the IWA Inspire Wales Award for Science and Technology in 2013. He also received Royal Recognition for his contributions to the British scientific community in October 2006.

He has co-authored two books; the first entitled "Practical Performance Modelling" by Kluwer in 2001 and the second entitled: "IMS: Deployment and Development Perspective" by John Wiley & Sons in 2009. He also edited 18 books, and authored more than 200 papers in refereed journals and conferences and filed two patents. He is member of the editorial board of several international journals and acted as Guest Editor for four journal special issues. He is the co-inventor of MOSEL: The Modelling, Specification and Evaluation Language. He has already supervised over 20 successful Ph.D. projects. He has also delivered over 15 key-notes in major international conferences.

Since 2003, he secured over £300K of consultancy in major companies such as Lucent Technologies, Siemens, Telekom Malaysia and General Dynamics. He also acted as Expert Witness in the major mobile IPR court cases between Nokia–Interdigital and Nokia–HTC.

Al-Begain's research interests span over fundamental and technological themes. He is working on developing analytical and numerical models for evaluating the performance of wireless networks; in particular, the performance of mission critical systems over Voice-over-LTE 4G networks. This is a project which is backed by the world-leader General Dynamics test bed facility. He is also leading a project for modelling behavioural changes to discover early symptoms of Dementia using hidden Markov models. He also developed both routing protocols for emergency systems as well as generalized queueing models for evaluating the impact of "Propagated Failures" resulting from a disaster on systems.

On the technological side, among many projects, he led the project on the IMS based next generation service and applications creation environment at ICRC supported by major industries with a value exceeding £1 million. When created, this facility was one of very few in universities worldwide. A recent consultancy within this theme resulted in the development of world-first Mission Critical Emergency System over VoLTE.

Al-Begain has led the bidding process and secured £6.4 million funding for establishing the Centre of Excellence in Mobile Applications and Services (CEMAS). CEMAS mission was to help SMEs to design, develop, commercialize as well as adopt cutting-edge and innovative mobile applications and technologies.

He was also the Principal Investigator on ICE-WISH EU FP7 project (€4.9 million total, €300K for UoG) with 17 partners from 11 countries.

Al-Begain has also a number of company directorships including being the CEO of Myliveguard Limited (http://www.myliveguard.com/), a spinout company of the University of Glamorgan exploiting an innovative Video-to-Mobile-Phone home/office security system based on his invention.

Chapter 1
ECMS: Its Position in the European Simulation History

Eugene J.H. Kerckhoffs

Abstract This contribution describes some major aspects of the history of European simulation conferences, from the early 1970s of the previous century on. The emphasis is on how the current "European Council for Modelling and Simulation" (ECMS) with its yearly ECMS-conferences has originated in this framework and continued up to now.

Keywords History · European simulation conferences · European Council of Modelling and Simulation (ECMS) · ECMS-Conferences

ECMS stands for "European Council for Modelling and Simulation", but also for its yearly conference "European Conference on Modelling and Simulation". A bit confusing; which meaning is meant should be clear in the context. When we speak, however, of "30 years ECMS", on the occasion of which this book is published, we point to ECMS 2016 as conference number 30 in the series of yearly conferences. The first conference in the series has been held in 1987, albeit by another name: ESM, European Simulation Multiconference. ECMS as an organization, i.e. as a Council, was established a few years later in 1990, but again by a different name: initially Continental Europe Simulation Council, but already after one year renamed ESC, European Simulation Council, a Council of SCS (the Society for Computer Simulation). The author was happy to be the initiator and first chair of this Council. In 2004, the Council changed its name into the current one: ECMS. From then on ECMS is an independent forum of European academics and professional users, dedicated to research, development and applications of modelling and simulation.

1st President of ECMS (formerly ESC), current ECMS Historian
Eugene J.H.Kerckhoffs-Retired

E.J.H. Kerckhoffs (✉)
Faculty of Electrical Engineering, Mathematics and Informatics, Delft University of Technology, Leiden, The Netherlands
e-mail: kerck037@planet.nl

K. Al-Begain and A. Bargiela (eds.), *Seminal Contributions to Modelling and Simulation*, Simulation Foundations, Methods and Applications,
DOI 10.1007/978-3-319-33786-9_1

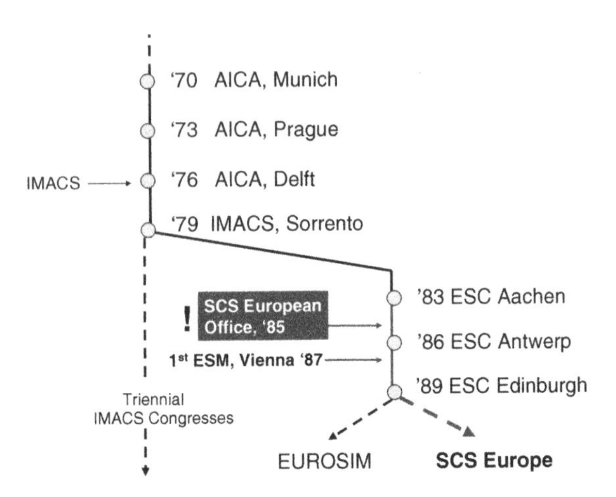

Fig. 1.1 History preceding ECMS

In the next paragraphs, we first pay attention to the simulation history before ECMS. What were the former circumstances in the European simulation community, which finally have led to the foundation of ECMS (formerly ESC)? This previous history is schematically represented in Fig. 1.1, where we start our overview in the early 1970s of the previous century.

In those days simulation was mainly continuous simulation, i.e. simulation of continuous systems; analogue and hybrid computers were used to run the simulations. For the European simulationists the major international Society was AICA, the World Association for Analogue Computing (established in 1956 in Brussels), with its triennial AICA World Congress. There existed also regional societies, such as for instance the Scandinavian Analogue Computer Society, founded in 1959 (and in 1968 renamed SIMS, Scandinavian Simulation Society). The first international congress, the author personally attended in his scientific life, was the AICA World Congress 1970 in Munich (Germany). Having been the secretary of both events, he will never forget the AICA World Congresses 1976 (Delft, the Netherlands) and 1979 (Sorrento, Italy), since both congresses have played a crucial role, which ultimately also has led to the birth of our Council ECMS (or better its precursor ESC).

During its world congress in Delft (1976), AICA was renamed IMACS (International Association for Mathematics and Computers in Simulation) to express that simulation, the covered field of interest, includes so much more than just continuous simulation and analogue/hybrid computing. From 1979 on the series of triennial IMACS World Congresses continued up to now. For especially the European simulation community the IMACS World Congress in Sorrento (1979) has been crucial. It was there that some European key simulationists get together and agreed to establish in the coming years new European "common-language based" Simulation Societies and to organize, apart from IMACS, their own triennial European Simulation Congress under the patronage of these new societies.

The first ESC (European Simulation Congress) was held in 1983 in Aachen (Germany) and organized by ASIM (the German speaking Simulation Society, founded in 1982) in close cooperation with DBSS (the Dutch speaking Simulation Society, founded in 1979). The 2nd ESC has been held in Antwerp (Belgium, 1986) and the 3rd one in Edinburgh (Scotland, 1989). During the last congress Eurosim (the Federation of European Simulation Societies) was established as an umbrella of the then consisting European Simulation Societies, such as the afore-mentioned ASIM and DBSS, the Italian Simulation Society ISCS (founded in 1985), etc.

In this European framework suddenly, in the mid-1980s, another player came on the European floor: SCS, the Society for Computer Simulation, an American Simulation Society, founded in 1954 in San Diego, with a Council in the UK (UKSC, founded in 1969). They (SCS) established in 1985 the SCS European Office at the University of Ghent in Belgium. This initiative must be seen as the very origin of the later series of ECMS Conferences!

Two years later, in 1987, the SCS European Office in Ghent organized, on behalf of the European part of SCS, an SCS European conference, actually the very first conference in our current series of 30 ECMS-Conferences! This conference has taken place in Vienna (Austria) with the name ESM (European Simulation Multiconference). Although originally the cooperation and understanding between SCS Europe and the European Simulation Societies of the Eurosim group was excellent—the 2nd European Simulation Congress ESC 1986 in Antwerp even was a common activity–, after the 3rd ESC in Edinburgh 1989 we see both organizations going their own way: the triennial European Simulation Congress ESC was continued as triennial congress, however now with the name “Eurosim Congress”, and SCS Europe continued its yearly ESM (European Simulation Multiconference).

In order to provide a forum to the European SCS members outside the UK and a scientific body behind its European ESM conferences, SCS decided, by decision of the Board of Directors in its annual meeting on July 18 1990 in Calgary (Canada), to approve the submitted European proposal to establish a Continental Europe Simulation Council (CESC) as second European Council of SCS, in addition to its first European Council UKSC. All European SCS members outside the UK were automatically CESC member. The author was the initiator and chair of this new SCS Council. Already one year later CESC was reorganized to cover whole Europe including the UK, and was therefore renamed European Simulation Council (ESC); all European members of SCS were from then on ESC members. UKSC was cancelled as a separate SCS Council and became, renamed UKSim, an independent Simulation Society and as such a member society of Eurosim.

As said above, from 1989 on SCS Europe continued, independent from other European Simulation Societies, its yearly European Simulation Multiconference (ESM). In Fig. 1.2 the conferences of the first 25 years in this series are shown, from the first one in Vienna (1987) till the 25th jubilee conference in Krakow (2011). One may notice that the name of the conference has changed somewhere on the run. From 2005 on we do not speak anymore of ESM (European Simulation Multiconference) but of ECMS (European Conference on Modelling and Simulation). By this name the conference is currently known. The reason of the change of name is a crucial change in the organization behind the conference. Until 2004 the SCS European Simulation Council and the SCS Europe Ltd (founded in 1995 to replace the SCS European Office) were, respectively, the organizational and financial body behind the conference. In 2004 however, during the 18th conference in Magdeburg, it was decided to cancel this dual system of responsibility and to continue with one organization ECMS (European Council for Modelling and Simulation), taking full responsibility for all organizational, scientific and financial issues with respect to its yearly conference; the conference was then renamed ECMS, European Conference on Modelling and Simulation.

The biggest conferences in the list of Fig. 1.2 were ESM 1990 (Nuremberg), ESM 1991 (Copenhagen) and ECMS 2007 (Prague), with each more than 200 attendants. The conferences were all in different locations, with one exception: Prague, where the conference has been held three times! All conferences were held somewhere in Europe, with again one exception: ECMS 2010 has been held in Kuala Lumpur (Malaysia), organized however by a European University (the University of Nottingham, Malaysia Campus).

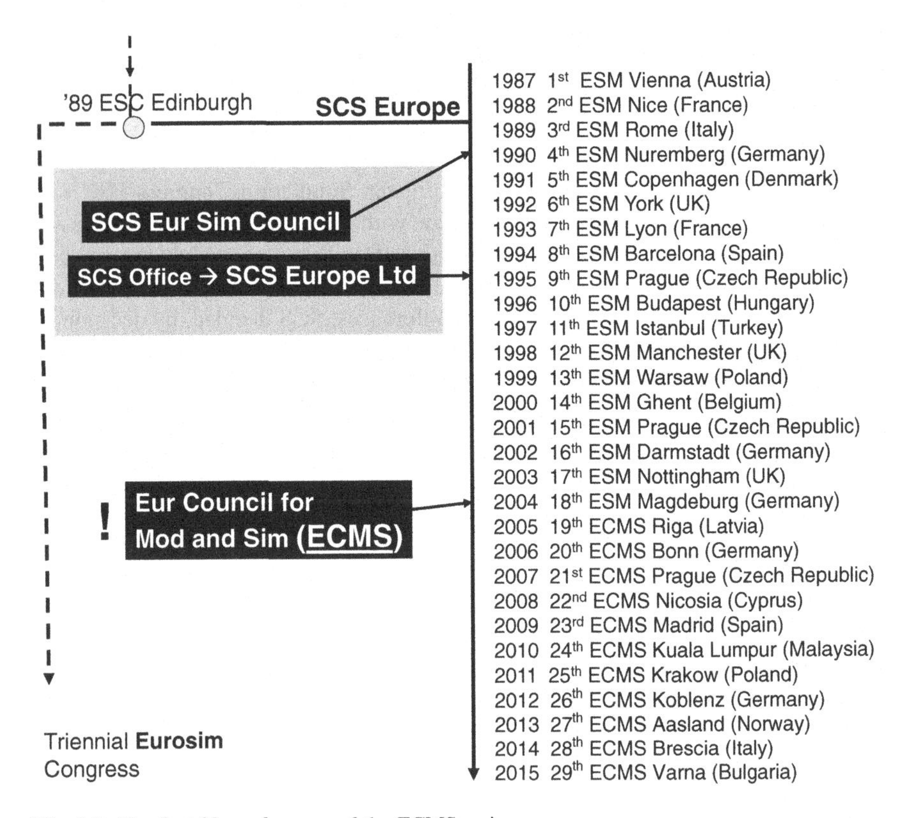

Fig. 1.2 The first 29 conferences of the ECMS series

The ECMS-Conferences of the last 5 years were: ECMS 2012 in Koblenz (Germany), ECMS 2013 in Aalesund (Norway), ECMS 2014 in Brescia (Italy), ECMS 2015 in Albena (near Varna, Bulgaria), and finally number 30 in the series: ECMS 2016 in Regensburg (Germany). All these conferences have been held in university locations, except for ECMS 2015 which has taken place in a well-equipped 5-stars hotel. The ECMS 2013 in Aalesund was special because of its very good organization and facilities; one of the keynote speakers there, prof. May-Britt Moser, got one year later in 2014 the Nobel Prize for Medicine.

ECMS is managed by a Board, chaired by the ECMS president. The presidents of ECMS (formerly ESC) were in chronological sequence: Eugene Kerckhoffs (the Netherlands, 1990–1997), Alexander Verbraeck (the Netherlands, 1997–2002), Andrzej Bargiela (UK, 2002–2006; during his presidency ESC changed into ECMS), Khalid Al-Begain (UK, 2006–2010), Andrzej Bargiela (2010–2012), Evtim Peytchev (UK, 2012–2014), and finally the current President Khalid Al-Begain (from 2014 on). Of course, the organization of the yearly ECMS Conference is not possible without the help of a "Council Office" to manage the reviewing process of the submitted papers, to manage the editing of the Proceedings, and last but not least to be the host during the conference and take care of the correct registration of the attendants. ECMS is lucky to have found in Martina-Maria Seidel a person to run the Office in an excellent manner. She started as assistant manager of the Office in 2002 (the 16th Conference, held in Darmstadt, Germany) and became a few years later the Office manager. She thus has taken care of 15 of the 30 conferences in the series; that is "1-in-2"!

The author would like to take this opportunity to congratulate ECMS with reaching this milestone of 30 conferences, and to thank all the volunteers and conference participants in these 30 years for their highly appreciated work and interest.

Chapter 2
Brain Maps for Space

May-Britt Moser

Abstract The brain controls spatial navigation in mammals by activating functionally specialized cell types in the medial temporal lobe. Key components of the spatial mapping system are place cells and grid cells. It has been known for some time that place cells are located in the hippocampus and are active only when the animal is entering a specific location in the environment. Here, we present our research results relating to grid cells. We found that grid cells are located upstream of the hippocampus, in the medial entorhinal cortex, and are activated whenever an animal enters locations that are distributed in a spatially periodic pattern across the environment. Moreover, we discovered that the grid cell network is intrinsically organized as grid cells clustered in distinct and independent grid maps with distinct scales, orientations and asymmetries, as well as distinct grid patterns of temporal organization.

Keywords Place cells · Grid cells · Grid maps · Spatial navigation

2.1 Main Contribution

The brain controls spatial navigation in mammals by activating functionally specialized cell types in the medial temporal lobe. A key component of the spatial mapping system is the place cell,located in the hippocampus. These cells—discovered by O'Keefe and Dostrovsky in 1971—are active only when the animal is entering a specific location in the environment. I will describe the discovery of another component of the mammalian spatial mapping system—the grid cell—which we found upstream of the hippocampus, in the medial entorhinal cortex, in 2005. Grid cells are activated

27th European Conference on Modelling and Simulation (ECMS) Ålesund, Norway, 27th May 2013.

M.-B. Moser (✉)
Norwegian University of Science and Technology, Kavli Institute for Systems Neuroscience and Centre for Neural Computation, Trondheim, Norway
e-mail: may-britt.moser@ntnu.no

K. Al-Begain and A. Bargiela (eds.), *Seminal Contributions to Modelling and Simulation*, Simulation Foundations, Methods and Applications,
DOI 10.1007/978-3-319-33786-9_2

whenever an animal enters locations that are distributed in a spatially periodic pattern across the environment. The repeating unit of the grid pattern much is an equilateral triangle. Grid cells are co-localized with head direction cells and border cells, which contain information about the direction in which the animal is moving and the boundaries of the environment. Despite the discovery of several elements of the mammalian spatial map, the interaction between the components is poorly understood. We addressed this question first by using optogenetics together with electrophysiological recordings of cells in the entorhinal cortex. Hippocampal neurons were infected with an adeno-associated virus carrying genes for a peptide tag that can be visualized by fluorescent antibodies as well as the light-sensitive cation channel channelrhodopsin-2 (ChR2). The virus was engineered to enable retrograde transport through axons of cells with projections into the hippocampus. Infected entorhinal cells were detected by local flashes of light. Channel rhodopsin-expressing cells responded with a short and constant latency to the light. All cell types in the entorhinal cortex were found to respond to the light, suggesting that place signals may be generated in the hippocampus by convergence of signals from all these entorhinal cell types. In addition to discussing the transformation of entorhinal to hippocampal spatial signals, I will devote a part of my talk to asking how the grid-cell network is intrinsically organized. To address this question, we used multi-channel recording from a much larger number of cells than recorded ever before in individual animals. Grid cells were found to cluster into a small number of modules with distinct grid scales, grid orientations and grid asymmetries, as well as distinct patterns of temporal organization. The different modules responded independently to changes in the geometry of the environment.

The existence of distinct and independent modules or grid maps makes entorhinal maps different from the many other sensory cortices where functions tend to be more graded and continuous.

This is in agreement with the suggestion that the grid map is a product of self-organizing network dynamics rather than specificity in the input. Because the crystal-like structure of the grid pattern is generated within the brain, not depending on specific sensory inputs, we are confronted with a unique situation in which we, by trying to understand how the grid pattern is formed, may obtain insights into how patterns are formed in the mammalian cortex.

2.2 Editors' Comments

Professor May-Britt Moser is the Founding Director of Centre for Neural Computation and co-Director of the Kavlil Institute for Systems Neuroscience. PhD in neurophysiology, University of Oslo 1995.

She is interested in the neural basis of spatial location and spatial specifically and cognition more generally. Her work, conducted with Edvard Moser as a long-term collaborator, includes the discovery of grid cells in the entorhinal cortex, as well as several additional space-representing cell types in the same circuit.

Her group is beginning to unravel the functional organization of the grid-cell circuit as well as its contribution to memory formation in the hippocampus.

May-Britt Moser was a co-Founder of the Centre for the Biology of Memory, a Research Council-funded Centre of Excellence from 2003 to 2012, and has taken on the Directorship of the Centre for Neural Computation, with a life time from 2013 to 2022.

May-Britt Moser Won the Nobel Prize in Physiology or Medicine 2014. She shares the Nobel Prize with Edvard Moser and John O'Keefe.

Chapter 3
Coordinated Learning of Entorhinal Grid Cells and Hippocampal Place Cells: Space, Time, Attention and Oscillations

Stephen Grossberg

Abstract The GridPlaceMap neural model explains and simulates how both entorhinal grid cells and hippocampal place cells may develop as spatial categories in a hierarchy of self-organizing maps by detecting, learning and remembering the most frequent and energetic co-occurrences of their inputs. The model simulates challenging behavioural and neurobiological data by embodying several parsimonious neural designs: Similar ring attractor mechanisms process the linear and angular path integration inputs that drive map learning; the same self-organizing map mechanisms can learn grid cell and place cell receptive fields; and the learning of the dorsoventral organization of multiple spatial scale modules through medial entorhinal cortex to hippocampus may use mechanisms homologous to those for temporal learning ("time cells") through lateral entorhinal cortex to hippocampus ("neural relativity"). Top-down hippocampus-to-entorhinal attentional mechanisms stabilize map learning, simulate how hippocampal inactivation may disrupt grid cells and help to explain data about theta, beta and gamma oscillations.

Keywords Grid cells · Place cells · Self-organizing map · Spatial navigation · Attention · Adaptive timing

S. Grossberg (✉)
Center for Adaptive Systems, Graduate Program in Cognitive and Neural Systems,
Departments of Mathematics, Psychology, and Biomedical Engineering,
Boston University, Boston, MA 02215, USA
e-mail: steve@bu.edu
URL: http://cns.bu.edu/~steve

K. Al-Begain and A. Bargiela (eds.), *Seminal Contributions to Modelling and Simulation*, Simulation Foundations, Methods and Applications,
DOI 10.1007/978-3-319-33786-9_3

3.1 How Do Grid Cells and Place Cells Arise Through Development and Learning?

Medial entorhinal grid cells and hippocampal place cells are crucial elements in brain systems for spatial navigation and episodic memory. This extended abstract summarizes the development of the GridPlaceMap neural model that explains many data about these cells and behaviours in a unified way. The articles in the reference list summarize and simulate these data. The current abstract provides a high-level overview of the results, along with the unifying neural design principles and mechanisms that enable them to be realized.

A place cell typically fires whenever an animal is present in one or more spatial regions, or places, of an environment. A grid cell typically fires in multiple spatial regions that form a regular hexagonal grid structure extending throughout an open-field environment. Different grid and place cells prefer spatially offset regions, with their firing fields increasing in size along the dorsoventral axes of the medial entorhinal cortex and hippocampus. The spacing between neighbouring fields for a grid cell also increases along the dorsoventral axis.

The GridPlaceMap neural model shows how grid cells and place cells may develop in a hierarchy of self-organizing maps. In this conception, grid cells and place cells are learned spatial categories in these maps. The model responds to realistic rat navigational trajectories by learning grid cells with hexagonal grid firing fields of multiple spatial scales, and place cells with one or more firing fields that match neurophysiological data about these cells and their development in juvenile rats. The place cells can represent much larger spaces than the grid cells, up to the least common multiple of the grid cell scales that activate each of them, thereby enabling the place cells to support navigational behaviours.

3.2 Homologous Self-organizing Map Laws for Grid Cell and Place Cell Learning

The model grid and place cells self-organizing maps both obey the same laws, and both amplify and learn to categorize the most frequent and energetic co-occurrences of their inputs. The different receptive field properties emerge because they experience different input sources. The place cells learn from the developing grid cells of multiple scales that input to them. The grid cells learn from stripe cells of multiple scales that input to them, with each stripe cell maximally sensitive to a different direction. The name "stripe cell" acknowledges that the spatial firing pattern of each such cell exhibits parallel stripes as the environment is navigated. Burgess and his colleagues have introduced an analogous concept of "band cells", but they are formed by the mechanism of oscillatory interference.

Grid and place cell learning occur equally well in GridPlaceMap models that are built up from either rate-based or spiking neurons. The results using spiking neurons build upon a previous rate-based model of grid and place cell learning, and thus illustrate a general method for converting rate-based adaptive neural models into models whose cells obey spiking dynamics. Remarkably, the spiking model continues to exhibit key analog properties of the data. New properties also arise in the spiking model, including the appearance of theta band modulation. The spiking model also opens a path for implementation in brain-emulating nanochips comprised of networks of noisy spiking neurons for learning to control autonomous adaptive robots capable of spatial navigation.

3.3 Learning the Dorsoventral Gradient of Receptive Field Sizes and Oscillation Frequencies

Both the spatial and temporal properties of grid cells vary along the dorsoventral axis of the medial entorhinal cortex. In vitro recordings of medial entorhinal layer II stellate cells have revealed subthreshold membrane potential oscillations (MPOs) whose temporal periods, and time constants of excitatory postsynaptic potentials (EPSPs), both increase along this axis. Slower (faster) subthreshold MPOs and slower (faster) EPSPs correlate with larger (smaller) grid spacings and field widths. The self-organizing map model simulates how the anatomical gradient of grid spatial scales can be learned by cells that respond more slowly along the gradient to their inputs from stripe cells of multiple scales. The model cells also exhibit MPO frequencies that covary with their response rates, and exhibit some properties of modular organization of the different spatial scales. The gradient in intrinsic rhythmicity is thus not compelling evidence for oscillatory interference as a mechanism of grid cell firing.

3.4 Homologous Spatial and Temporal Mechanisms: Neural Relativity

This spatial gradient mechanism is homologous to a gradient mechanism for temporal learning in the lateral entorhinal cortex and its hippocampal projections that was proposed in the 1980s. Such adaptively timed learning has simulated data about the role of hippocampus in supporting learning that bridges temporal gaps, such as occurs during trace conditioning and delayed matching-to-sample. This type of "spectrally timed learning" has Weber Law properties that have been confirmed by recent experiments that have discovered "time cells" in the hippocampus. Spatial and temporal representations may hereby arise from homologous mechanisms, thereby embodying a mechanistic "neural relativity" that may clarify how episodic memories are learned.

3.5 Homologous Processing of Angular and Linear Velocity Path Integration Inputs

The inputs that drive the initial development of grid cells and place cells are angular and linear velocity signals that are activated by an animal's navigational movements. The model proposes that both angular and linear velocity signals are processed by ring attractor neural circuits. Angular velocity signals are proposed to be processed by ring attractors that are composed of head direction cells, whereas linear velocity signals are proposed to be processed by ring attractors that are composed of stripe cells. Stripe cells were predicted during the development of the model and have received neurophysiological support since that time.

The outputs of head direction cells modulate the linear velocity signals to multiple directionally-selective stripe cell ring attractor circuits. This modulation is sensitive to the cosine of the difference between the current heading direction of movement and the ring attractor's directional preference. Each stripe cell ring attractor is sensitive to a different direction and spatial scale. Stripe cells are the individual cells within each such ring attractor circuit and are activated at different spatial phases as the activity bump moves across their ring locations. They may be activated periodically as the activity bump moves around the ring attractor more than once in response to the navigational movements of the animal.

The model's assumption that both head direction cells and stripe cells are computed by ring attractors that drive grid and place cell development is consistent with data showing that adult-like head direction cells already exist in parahippocampal regions of rat pups when they actively move out of their nests for the first time at around two weeks of age.

3.6 Stable Learning, Attention, Realignment and Remapping

Place cell selectivity can develop within seconds to minutes, and can remain stable for months. The hippocampus needs additional mechanisms to ensure this long-term stability. This combination of fast learning and stable memory is often called the *stability–plasticity dilemma*. Self-organizing maps are themselves insufficient to solve the stability–plasticity dilemma in environments whose input patterns are dense and are nonstationary through time, as occurs regularly during real-world navigation. However, self-organizing maps augmented by learned top-down expectations that focus attention upon expected combinations of features can dynamically stabilize their learned categories, whether they are the spatial categories that are needed during navigation, or the cognitive categories that are needed during object and scene recognition.

Adaptive Resonance Theory, or ART, has predicted since 1976 how to dynamically stabilize the learned categorical memories of self-organizing maps by using learned top-down expectations and attentional focusing. Experimental data about the hippocampus from several labs are compatible with the predicted role of top-down expectations and attentional matching in memory stabilization. These experiments clarify how cognitive processes like attention may play a role in entorhinal-hippocampal spatial learning and memory stability. The proposed mechanism of top-down attentional matching may also help to clarify data about grid and place cell remapping and alignment of grid orientations.

3.7 Beta, Gamma and Theta Oscillations

Within ART, a sufficiently good match of a top-down expectation with a bottom-up input pattern can trigger fast gamma oscillations that enable spike-timing-dependent plasticity to occur. In contrast, a big enough mismatch can trigger slow beta oscillations that do not. Such beta oscillations have been reported to occur in the hippocampus during the learning of novel place cells, and have the properties expected when mismatches occur and receptive field refinements are learned. In particular, there seems to be an inverted-U in beta power through time as a rat navigates a novel environment. Beta oscillations also occur at the expected times in visual cortex and in the frontal eye fields during shifts in spatial attention. Thus, the match/mismatch dynamics leading to gamma/beta oscillations seem to occur in multiple brain systems.

The theta rhythm has been associated with properties of spatial navigation, as has firing of entorhinal grid cells. Recent experiments have reduced the theta rhythm by inactivating the medial septum (MS) and demonstrated a correlated reduction in the hexagonal spatial firing patterns of grid cells. These results, along with properties of intrinsic membrane potential oscillations in slice preparations of entorhinal cells, have been proposed to support an oscillatory interference model of grid cells. Our self-organizing map model of grid cells can, however, explain these data as emergent properties of model dynamics, without invoking oscillatory interference. In particular, the adverse effects of MS inactivation on grid cells can be understood from how the concomitant reduction in cholinergic inputs may increase conductances of leak potassium and slow and medium after-hyperpolarization channels.

3.8 Model Parsimony

Our emerging neural theory of spatial and temporal processing in the entorhinal-hippocampal system exhibits a remarkable parsimony and unity in at least three ways: It proposes that similar ring attractor mechanisms compute the linear and angular path integration inputs that drive map learning; that the same self-organizing map mechanisms can learn grid cell and place cell receptive fields, despite their dramatically different appearances; and that that the dorsoventral gradient of multiple scales and modules of spatial learning through the medial entorhinal cortex to hippocampus may use mechanisms that are homologous to mechanisms earlier proposed for temporal learning through the lateral entorhinal cortex to hippocampus ("neural relativity"), as reflected by data about trace conditioning, delayed matching-to-sample, and "time cells." This mechanistic homolog clarifies why both spatial and temporal processing occur in the entorhinal–hippocampal system and why episodic learning may be supported by this system. No less striking is the fact that both grid cells and place cells can develop by detecting, learning and remembering the most frequent and energetic co-occurrences of their inputs, properly understood. This co-occurrence property is naturally computed in response to data, such as navigational signals, that take on contextual meaning through time.

Acknowledgement This research is supported in part by the SyNAPSE program of DARPA (HR0011-09-C-0001).

Bibliography

1. Fortenberry B, Gorchetchnikov A, Grossberg S (2012) Learned integration of visual, vestibular, and motor cues in multiple brain regions computes head direction during visually-guided navigation. Hippocampus 22:2219–2237
2. Gorchetchnikov A, Grossberg S (2007) Space, time, and learning in the hippocampus: how fine spatial and temporal scales are expanded into population codes for behavioral control. Neural Netw 20:182–193
3. Grossberg S (2009) Beta oscillations and hippocampal place cell learning during exploration of novel environments. Hippocampus 19:881–885
4. Grossberg S (2013) Adaptive resonance theory: how a brain learns to consciously attend, learn, and recognize a changing world. Neural Netw 37:1–47
5. Grossberg S, Merrill JWL (1992) A neural network model of adaptively timed reinforcement learning and hippocampal dynamics. Cogn Brain Res 1:3–38
6. Grossberg S, Merrill JWL (1996). The hippocampus and cerebellum in adaptively timed learning, recognition, and movement. J Cogn Neurosci 8:257–277
7. Grossberg S, Pilly PK (2012) How entorhinal grid cells may learn multiple spatial scales from a dorsoventral gradient of cell response rates in a self-organizing map. PLoS Comput Biol 8 (10):31002648. doi:10.1371/journal.pcbi.1002648
8. Grossberg S, Pilly PK (2014) Coordinated learning of grid cell and place cell spatial and temporal properties: multiple scales, and oscillations. Philos Trans R Soc B. 369:20120524
9. Grossberg S, Versace M (2008) Spikes, synchrony, and attentive learning by laminar thalamocortical circuits. Brain Res 1218:278–312
10. Grossberg S, Schmajuk NA (1989) Neural dynamics of adaptive timing and temporal discrimination during associative learning. Neural Netw 2:79–102
11. Mhatre H, Gorchetchnikov A, Grossberg S (2012) Grid cell hexagonal patterns formed by fast self-organized learning within entorhinal cortex. Hippocampus 22:320–334
12. Pilly PK, Grossberg S (2012) How do spatial learning and memory occur in the brain? Coordinated learning of entorhinal grid cells and hippocampal place cells. J Cogn Neurosci 24:1031–1054

13. Pilly PK, Grossberg S (2013) How reduction of theta rhythm by medium septum inactivation may disrupt periodic spatial responses of entorhinal grid cells by reduced cholinergic transmission. Frontiers Neural Circu. doi:10.3389/fncir.l2013.00173
14. Pilly PK, Grossberg S (2013) Spiking neurons in a hierarchical model can learn to develop spatial and temporal properties of entorhinal grid cells and hippocampal place cells. PLoS ONE, http://dx.plos.org/10.1371/journal.pone.0060599
15. Pilly PK, Grossberg S (2014) How does the modular organization of entorhinal grid cells develop? Frontiers Human Neurosci. doi:10.3389/fnhum.2014.0037

Chapter 4
Adaptive Model Theory: Modelling the Modeller

Peter D. Neilson and Megan D. Neilson

Abstract Adaptive Model Theory is a computational theory of the brain processes that control purposive coordinated human movement. It sets out a feedforward-feedback optimal control system that employs both forward and inverse adaptive models of (i) muscles and their reflex systems, (ii) biomechanical loads on muscles, and (iii) the external world with which the body interacts. From a computational perspective, formation of these adaptive models presents a major challenge. All three systems are high dimensional, multiple input, multiple output, redundant, time-varying, nonlinear and dynamic. The use of Volterra or Wiener kernel modelling is prohibited because the resulting huge number of parameters is not feasible in a neural implementation. Nevertheless, it is well demonstrated behaviourally that the nervous system does form adaptive models of these systems that are memorized, selected and switched according to task. Adaptive Model Theory describes biologically realistic processes using neural adaptive filters that provide solutions to the above modelling challenges. In so doing we seek to model the supreme modeller that is the human brain.

Keywords Adaptive nonlinear models · Neural adaptive filters · Feature extraction · Movement synergies · Riemannian geometry

The human brain is an analogical modelling device. It forms adaptive models of the environment and of the body in interaction with the environment and it uses these models in the planning and control of purposive movement. The movement system includes the entire musculoskeletal system in interaction with the environment. There are some 700 functional muscles (groups of muscle fibres with the same mechanical action that can be controlled independently by the nervous system)

P.D. Neilson (✉) · M.D. Neilson
School of Electrical Engineering & Telecommunications,
University of New South Wales, Sydney 2052, Australia
e-mail: p.neilson@unsw.edu.au

M.D. Neilson
e-mail: megan.neilson@gmail.com

K. Al-Begain and A. Bargiela (eds.), *Seminal Contributions to Modelling and Simulation*, Simulation Foundations, Methods and Applications,
DOI 10.1007/978-3-319-33786-9_4

controlling about 110 elemental movements. From the perspective of the brain, the system to be controlled consists of three multiple input–multiple output, redundant, time-varying, nonlinear dynamical systems connected in cascade (i) the muscle control system (muscles and their reflex systems), (ii) the biomechanical system (biomechanical loads on muscles), and (iii) the external system (external world). Sensory systems continuously monitor the input and output signals of all three of these subsystems and form adaptive models of the nonlinear dynamical relations within and between the various sensory modalities involved. The brain compares model predictions with actual sensory signals (afference) and takes discrepancies very seriously. Discrepancies lead to an increase in brain activity as the brain analyses errors and attempts to update its models. It also defends against perturbations by slowing movements and stiffening.

While it can be demonstrated behaviourally that the nervous system forms these adaptive models, many technical issues stand in the way of understanding computationally how it does so. Not least of these technical issues is the large amount of redundancy in afferent signals. Tens of thousands of receptors are involved in detecting, for example, muscle tensions and elemental movements. It is not possible to form accurate models of multiple input–multiple output systems when the inputs are strongly interrelated, and it is not possible to form accurate inverse models when the number of outputs is less than the number of inputs or when the outputs and/or inputs are contaminated with large amounts of noise. Because of this the nervous system has to remove redundancy from afferent signals and extract well-conditioned independently varying (orthogonal) signals suitable for adaptive modelling.

It does this in two stages involving (i) slow adaptation and (ii) fast adaptation, respectively. (i) Slow adaptation involves correlational-based mechanisms of synaptic plasticity in networks of neurons connecting sensory receptors to the cerebral cortex (sensory pathways). Over several weeks to months, correlational-based mechanisms of synaptic plasticity lead to the adaptive formation of sensory maps in the cortex. In the somatosensory cortex, for example, maps of elemental movements and of the lengths and tensions of functional muscles are formed. Similar sensory maps for all other sensory modalities are formed in other cortical areas of the brain. For example, cortical columns in the primary visual cortex adaptively tune their synaptic weights to extract stochastic, orthogonal, visual features from cortical receptive fields on the left and right retinas.

(ii) According to Adaptive Model Theory, fast adaptation depends on the existence of networks of neural adaptive filters within every sensory modality. The tuneable input-output transfer characteristic of each neural adaptive filter is preset in anticipation of a planned movement by patterns of neural activity held on-line in working memory. This activity modulates the synaptic gains of neurons in the adaptive filters. The modulating activity can be transferred from working memory into intermediate and long-term memory and, given an appropriate memory selection code, can be retrieved from intermediate or long-term memory and reinstated as a pattern of neural activity in working memory. This enables the tuning of neural adaptive filters to be switched quickly and smoothly between a library of settings acquired through previous sensory-motor experience and stored in memory.

The network of neural adaptive filters within each sensory modality extracts a small number of independently varying feature signals (determined by the movement being performed) from the large number of covarying signals within the sensory map for that modality. It does this by performing a type of Gramm-Schmidt orthogonalization (or QR factorization) but, instead of algebraic regression coefficients as used in these algorithms, it employs neural adaptive filters with nonlinear dynamical transfer characteristics. For example, when turning the steering wheel of a car some 20 elemental movements at the shoulders, elbows, forearms and wrists have to be coordinated to vary together in a nonlinear dynamically coupled way in order to generate a single movement feature signal corresponding to rotation of the steering wheel. The network of neural adaptive filters "learns" this coordination by adaptively modelling the nonlinear dynamical relationships between the covarying elemental movement signals in the somatosensory cortical sensory map and by extracting a single independently varying movement feature signal, the controlled degree of freedom of the task. Higher level planning and control of the movement can then be carried out within this low-dimensional subspace thereby greatly reducing demand on central processing resources. Slave copies of the same neural adaptive filters can be employed in the inverse direction to transform low-dimensional required feature signals planned centrally back into a large number of covarying elemental movements.

Similar feature extraction takes place simultaneously in every sensory modality. The resulting feature signals in each modality are then well-conditioned for forming adaptive models of the nonlinear dynamical forward and inverse relationships between them. These between-modality models are used in the planning and control of movement. For example, in steering a car, the required visual response associated with steering the car can be transformed into the elemental movements required to turn the steering wheel, thence into the muscle tensions that must be generated to produce those movements, and finally into the descending motor commands required to generate those tensions. Because there are many more descending motor commands than there are functional muscles, and many more functional muscles than there are elemental movements, and many more elemental movements than required to perform the task, a problem of redundancy exists in the formation and selection of appropriate models for a task. Many different sets of orthogonal feature signals and dynamical relationships between them can be used for any given task. Using methods of Riemannian geometry to take into account the nonlinear inertial interactions within and between the body and the environment, a unique set of orthogonal feature signals (and the dynamical relationships between them) can be selected to achieve task goals with minimal muscular effort.

Modellers of multi-variable, redundant, time-varying, nonlinear dynamical systems are confronted by the very large numbers of parameters required to model such systems. In the case of the adaptive models described above the tens of thousands of parameters required to form Volterra or Wiener kernel models render such models impractical within the nervous system. In Adaptive Model Theory we explore circuitry in cortico-cerebellar-cortical pathways in the brain capable of functioning as neural adaptive filters with third-order nonlinear dynamical transfer characteristics that require only 60 parameters for each single input—single output nonlinear relationship. We hold that the theory demonstrates a means by which such circuitry can model the relationships between the incoming sensory signals despite these typically being non-Gaussian, non-white and non-stationary. In so doing, the theory delineates a neurally realistic account of how the adaptive models necessary for movement control can be achieved. In that sense Adaptive Model Theory seeks to model the supreme modeller that is the human brain. A complete account of this approach and its background is available in publications [1–9].

References

1. Neilson PD, Neilson MD, O'Dwyer NJ (1985) Acquisition of motor skills in tracking tasks: learning internal models. In: Russell DG, Abernethy B (eds) Motor memory and control: the otago symposium, Dunedin, New Zealand, 1982. Human Performance Associates, Dunedin, NZ, pp 25–36
2. Neilson PD, Neilson MD, O'Dwyer NJ (1988) Internal models and intermittency: a theoretical account of human tracking behavior. Biol Cybern 58:101–112
3. Neilson PD, Neilson MD, O'Dwyer NJ (1992) Adaptive model theory: application to disorders of motor control. In: Summers JJ (ed) Approaches to the study of motor control and learning. North Holland, Amsterdam, pp 495–548
4. Neilson PD (1993) The problem of redundancy in movement control: the adaptive model theory approach. Psychol Res 55:99–106
5. Neilson PD, Neilson MD, O'Dwyer NJ (1997) Adaptive model theory: central processing in acquisition of skill. In: Connolly K, Forssberg H (eds) Neurophysiology and neuropsychology of motor development. Mac Keith Press, London, pp 346–370
6. Neilson PD, Neilson MD (2005) An overview of adaptive model theory: solving the problems of redundancy, resources, and nonlinear interactions in human movement control. J Neural Eng 2:S279–S312
7. Bye RT, Neilson PD (2008) The BUMP model of response planning: variable horizon predictive control accounts for the speed–accuracy tradeoffs and velocity profiles of aimed movement. Hum Mov Sci 27:771–798
8. Neilson PD, Neilson MD (2010) On theory of motor synergies. Hum Mov Sci 29:655–683
9. Neilson PD, Neilson MD, Bye RT (2015) A Riemannian geometry theory of human movement: the geodesic synergy hypothesis. Hum Mov Sci 44:42–72

Chapter 5
Multi-formalism Modelling of Human Organization

Alexander H. Levis

Abstract The modelling of a human organization for the analysis of its behaviour in response to external stimuli is a complex problem and requires development and interoperation of a set of several models. Each model, developed using different modelling languages but the same data, offers unique insights and makes specific assumptions about the organization being modelled. Interoperation of such models can produce a more robust modelling and simulation capability to support analysis and evaluation of the organizational behaviour. The meta-modelling analysis based on concept maps and ontologies indicates what types of interoperation are valid between models expressed in different modelling languages. The approach is illustrated through several examples focusing on the use of social networks, Timed Influence Nets (a variant of Bayes Nets) and Colored Petri nets.

Keywords Timed influence nets · Social networks · Concept maps · Ontology · Workflow

5.1 Framing the Multi-formalism Modelling Problem

Modelling is the process of producing a model; a model is a representation of the construction and working of some situation of interest [20]. Figure 5.1 represents the modelling hierarchy where a model is obtained using a modelling tool that applies a modelling formalism or language to represent a specific situation. The model itself should always conform to the modelling formalism used to create it.

Multiple models are used because each modelling formalism provides certain capabilities and makes specific assumptions about the domain or situation being modelled. For example, Timed Influence Nets [9] describe cause and effect relationships among groups at high level, but have no capability of capturing social

A.H. Levis (✉)
Electrical and Computer Engineering, George Mason University,
Fairfax, VA 20105, USA
e-mail: alevis@gmu.edu

K. Al-Begain and A. Bargiela (eds.), *Seminal Contributions to Modelling and Simulation*, Simulation Foundations, Methods and Applications,
DOI 10.1007/978-3-319-33786-9_5

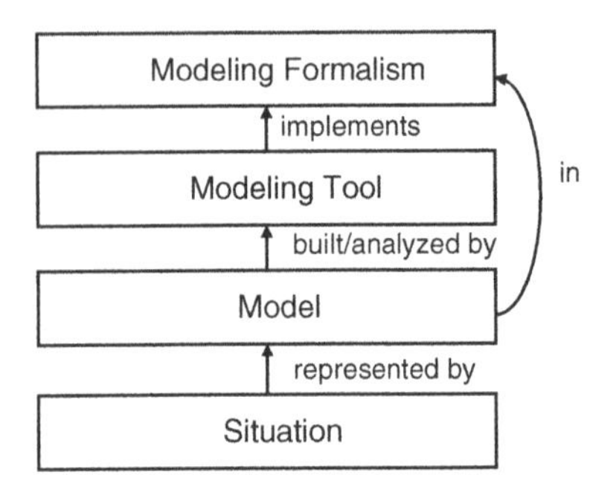

Fig. 5.1 Modelling hierarchy

aspects among the groups of interest. Social networks [4], on the other hand, can describe the interactions among groups and members of the groups. In this context, a multi-formalism modelling process addresses a complex problem through the use of a number of interconnected domain-specific models where each model contributes insights to the overall problem. The interoperations between the interconnected models could serve different purposes and can happen in various forms. The focus of this chapter is on modelling human organizations.

Social scientists have been collecting and analyzing sociocultural data to study social groups, organizations, tribes, urban and rural populations, ethnic groups and societies. Sometimes the data are longitudinal (i.e. time series) tracking a particular social entity over a long time period and other times the data are taken at a single point in time (e.g. surveys) across a diverse population. Sometimes the data are focused on individual actors and their attributes and other times on the relations between actors. This poses a unique challenge: the classification and taxonomy of model types and how data relate to them, i.e. a taxonomy that can bridge the gap between data and models. This challenge is further exacerbated by the fact that no single model can capture the complexities of human behaviour especially when interactions among groups or organizations with diverse social and cultural attributes are concerned.

Because the human behaviour domain is very complex when observed from different perspectives, a classification scheme is needed that can place different modelling approaches and modelling formalisms in context, but in a way that is meaningful to empiricists who collect the data, to modellers that need data to develop and test their models and to theoreticians who use model-generated data to induce theoretical insights. As a start to the development of a taxonomy for the problem space, four dimensions are considered.

Mathematical and Computational Modelling Languages. There are many modelling languages and many classifications of them, e.g. static or dynamic, deterministic or stochastic, linear or nonlinear, continuous time or discrete time, time driven or event driven, quantitative or qualitative, agent centric or system centric and so forth. On the basis of these modelling languages, different types of models have been developed: Bayesian networks, Social Networks, Dynamic Networks, Colored Petri Nets, State Machine models, Dynamics and Control models based on differential equations, System Dynamics models based on difference equations, Agent-Based models with internal sets of rules, Econometric models (time series based models), Statistical models, Input-Output (e.g. Leontieff) models and many others.

Social Entity (Granularity). This is an important dimension because it addresses the issue of improper generalization of the data. Being explicit in specifying the characterization of the social entity that will be modelled and analyzed is essential. The social entity classification can range from an individual (e.g. the leader of a country, or a chief executive officer, or a military leader) to a cell or team (e.g. an aircraft maintenance crew or a terrorist cell,) to a clan or tribe, to an ethnic or religious group, all the way to a multi-cultural population—the society—of a nation-state. There are serious definitional issues with regard to the social entity that need to be addressed by mathematical and computational modellers. These issues include spatial, temporal and boundary constraints; sphere of influence; change processes and rate of adaptation. For example, while the boundaries of an individual are impermeable, the boundaries of a tribe are permeable. Marriage enables crossing clan or tribal boundaries and, interestingly, in both directions. Individuals can be in only one location at a time; whereas, groups can have a very complex spatial-temporal presence (e.g. ranging from an agricultural cooperative to a transnational organization). As the granularity of the social entity increases, the size of the sphere of influence increases, the physical space covered increases and the rate of adaptation decreases. For example, individuals can impact those within their communication network and adapt to new situations in minutes whereas a nation-state can influence other nation-states but may take years to adapt. Finally, the change processes are different with individuals changing as they learn and nation-states changing through processes such as migration, legislation, economic collapse and war.

Scope of Problem. In the military domain, the types of decisions that are made are characterized as tactical, operational and strategic. It used to be that tactical decisions concerned the present and immediate future, operational decisions the near-term future, while strategic decisions concerned the long term (Fig. 5.2). However, information technology has changed this; it has made it possible for tactical decisions

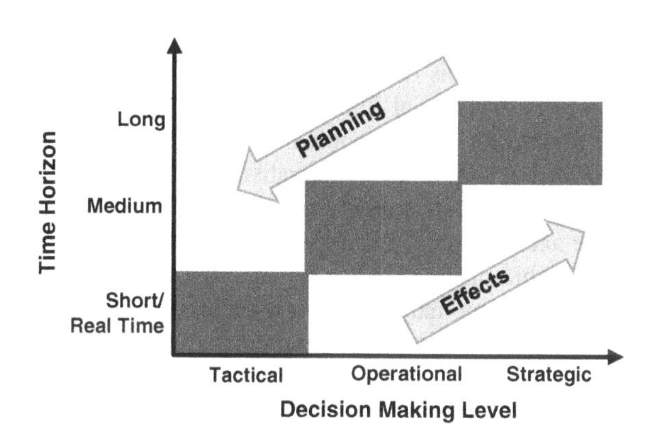

Fig. 5.2 Problem scope and time horizon

to have almost immediate strategic implications. The three levels have become very much coupled. For example, a terrorist act, itself a small tactical operation, can have significant strategic impact by changing the long-term behaviour of a population (e.g. the effect of an unsuccessful terrorist action on an airplane caused significant changes in air travel security worldwide.)

Time. The time attribute is complex and cannot be captured by a single variable. First, there is the time period that the data cover. This requires at least two variables: one is the location on the time line and the other is the duration of the period the data cover. Another possible attribute is the time interval between sampling instants. Second, there is the time dimension of the model itself. It can be an instant or a period.

The choice of an instant leads to static models; the instant of time can be a year, a month, a week, or any other epoch. Essentially, a static description of an organization is developed that remains unchanged during that epoch. The other alternative is to consider a time interval over which the organization transforms. Again, that time interval can range from hours to years. A key attribute of the time interval (duration) is its location on the timeline. Third, there is the time dimension that the model addresses. This is called the time horizon of the model.

The time dimension is closely related to the scope of the problem and is also related to the size of the social entity considered—as the time horizon is increased, the sphere of influence (i.e. the size of the social entity that will be impacted) is increased. Furthermore, the larger the social entity is, the longer the time horizon necessary to observe effects on the behaviour. While exceptions to this rule are possible, this consideration also restricts the combinations of entity size and time horizon that define meaningful problems.

These four types of dimensions (modelling formalism, social entity, scope and time) define the problem space for modelling the behaviour of diverse types of human organizations. In many cases, the mathematical and computational models are created so that they can be used in developing strategies to effect change that will increase the effectiveness of the organization. In this case, proper handling of time is essential both in the model as well as in the underlying data. An additional challenge with respect to the temporal dimension(s) is to consider the persistence of effects on the targeted social entity; it is important to account not only for delays in an effect becoming observable, but also for its gradual attenuation over time.

Aspects of this construct can be represented graphically in a three-dimensional space that has Social Entity, Time and Modelling Formalism or Language, being the three axes. In the graphic in Fig. 5.3 some possible calibrations of the axes are shown, but this is only illustrative. Social scientists and mathematical and computational modellers working together may refine the attributes that are marked on these axes and place them in a logical order. The scope axis has not been included; the graphic of Fig. 5.3 can be interpreted as being for a given scope. Indeed, the choice of scope may lead to a re-calibration of the three axes.

Given a classification scheme and the appropriate ordering, then we can define a set of feasible cells in that space. Each cell then defines a particular model by specifying three attributes: (a) the social entity being modelled, (b) the epoch being considered

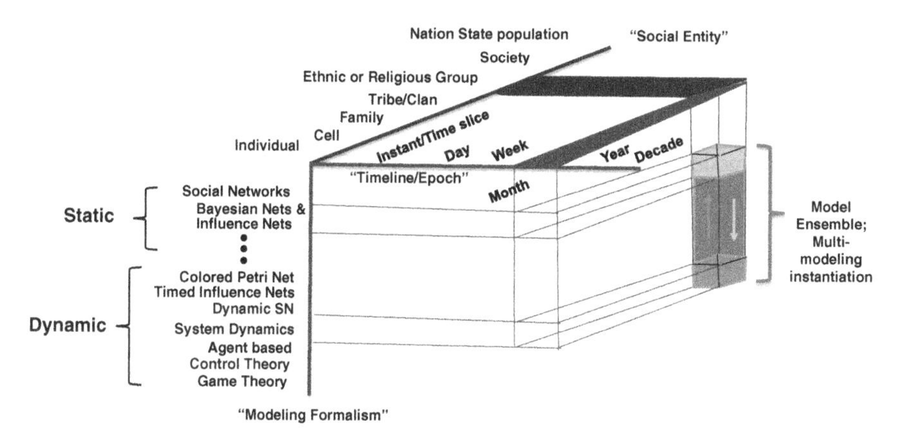

Fig. 5.3 A possible classification scheme for multi-formalism modelling

and (c) the modelling formalism or language. Not all cells in the space may be of interest in the sense that no problems addressing the situation of interest have been identified for that combination of attributes. Also, some mathematical and computational models, by their basic assumptions, may not be able to be associated with specific values of the other axes. For example, agent-based models are not appropriate for modelling a single individual so the corresponding set of cells would be empty. Within each cell there may be multiple models depending on the problem scope.

The classification scheme selected for each axis should make sense with respect to the problem/issue domain. A particular challenge faced earlier while modelling the dynamics of a province in Iraq with a set of externally provided data is that terrorist organizations exist and operate amidst civilian (non-combatant) populations. Since no single model can capture well both the terrorist cell dynamics and the non-combatant population behaviour, a set of interoperating models was needed. In another example, the impact of a course of action focused on de-escalating a crisis between two nation states with nuclear capability (e.g. Pakistan and India) needed to be assessed. An ensemble of models, each using different modelling formalism, was used to explore the impact of applying the de-escalatory strategy on the behaviour of the two governments.

5.2 Multi-formalism Modelling Challenges

The classification scheme of Fig. 5.3 would act as the first filter that would show which "cells" are available for possible use (i.e. which modelling formalisms) based on the scope, social entity and time dimension of the issue being considered and which ones are not appropriate for use because of granularity or temporal considerations.

Suppose now that a set of non-empty cells has been identified in Fig. 5.3 that could be used to address the problem of interest. The existence of multiple modelling alternatives leads to a set of questions: (a) What data are required for each modelling formalism and are that data available; (b) will the selected models run independently or will they interact or interoperate? (c) If they interoperate, what form will the interoperation take?

Note that model interactions can take a wide variety of forms: (1) Two models run in series with the output of one providing an input to the other; (2) One model runs inside another; (3) Two models run side-by-side and interoperate; and (4) One model is run/used to construct another by providing design parameters and constraints or to construct the whole or part of another model. The interoperation can be *complementary* where the two run totally independently of each other supplying parts of the solution required to answer the questions, or *supplementary* where the two supply (offline and/or online) each other with parameter values and/or functionality not available to either individual model; These are all aspects of the need for *semantic interoperability*.

In order to address in a structured manner the modelling and simulation issues that arise when multiple models are to interoperate, four layers need to be addressed (Fig. 5.4). The first layer, the *Physical Layer*, i.e. Hardware and Software, is a platform that enables the concurrent execution of multiple models expressed in different modelling languages and provides the ability to exchange data and also to schedule the events across the different models. The second layer is the *Syntactic Layer* which ascertains that the right data are exchanged among the models. Once this is achieved, a third problem needs to be addressed at the *Semantic Layer*, where the interoperation of different models is examined to ensure that conflicting assumption in different modelling languages are recognized and form constraints to the exchange of data. In the *Workflow Layer* valid combinations of interoperating models are considered to address specific issues. Different issues require different workflows. The use of multiple interoperating models is referred to as Multi-Formalism Modelling (MFM) while the analysis of the *validity* of model interoperation is referred to as Meta-Modelling. Each of these layers and the challenges that they pose are addressed in the next section.

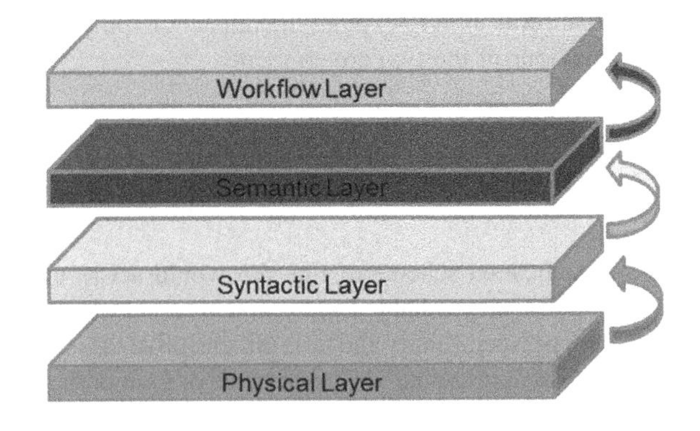

Fig. 5.4 The four layers of multi-formalism modelling

5.3 The Multi-formalism Modelling Layers

5.3.1 The Physical and Syntactic Layers

The technical issues regarding the physical and syntactic layers have been resolved a decade ago. There are numerous testbeds, based on different principles, which enable the interoperation of models. For example, SORASCS, developed at CASOS at Carnegie Mellon University used a Service-Oriented Architecture (SOA) [8]. The C2 Wind Tunnel (C2WT) on the other hand is an integrated, multi-modelling simulation environment [10, 13]. Its framework uses a discrete event model of computation as the common semantic framework for the precise integration of an extensible range of simulation engines, using the Run-Time Infrastructure (RTI) of the High-Level Architecture (HLA) platform. The C2WT offers a solution for multi-model simulation by decomposing the problem into model integration and experiment or simulation integration tasks as shown in Fig. 5.5.

Model Integration: Integrated experiments or simulations are specified by a suite of domain-specific models, including for instance: human organizations (expressed using the Colored Petri Net modelling language and Social Networks), computer and communication networks (OMNET++ network simulation language), physical platforms (Matlab/Simulink based models) and the physical environment (e.g. Google Earth). While the individual behaviours simulated by the different simulation models are essential, they must interact as specified by the workflow for the particular simulation or experiment. Their interactions need to be formally captured and the simulation of the components needs to be coordinated. This is a significant challenge, since the component models are defined using dramatically different domain-specific modelling languages. To achieve these objectives, the C2WT uses the meta-modelling technology and the Vanderbilt MIC tool suite.[1] The key new component is the model integration layer (Fig. 5.5), where a dedicated model integration language (MIL) is used for model integration. The MIL consists of a carefully selected collection of modelling concepts that represent the domain-specific simulation tools. An additional feature of the C2WT is its ability to include human-in-the-loop architectures whether in the roles of operators (piloting drones) or playing the roles of decision makers.

Model-based Experiment Integration: C2WT uses the MIC model interpretation infrastructure for the generators that automatically integrate heterogeneous

[1]MIC is a meta-programmable Model-Integrated Computing (MIC) tool suite for integrating models, to manage the configuration and deployment of scenarios in the simulation environment, and to generate the necessary interface code for each integrated simulation platform. It has evolved over two decades of research at the Institute for Software Integrated Systems at Vanderbilt University and is now used in a wide range of government and industry applications.

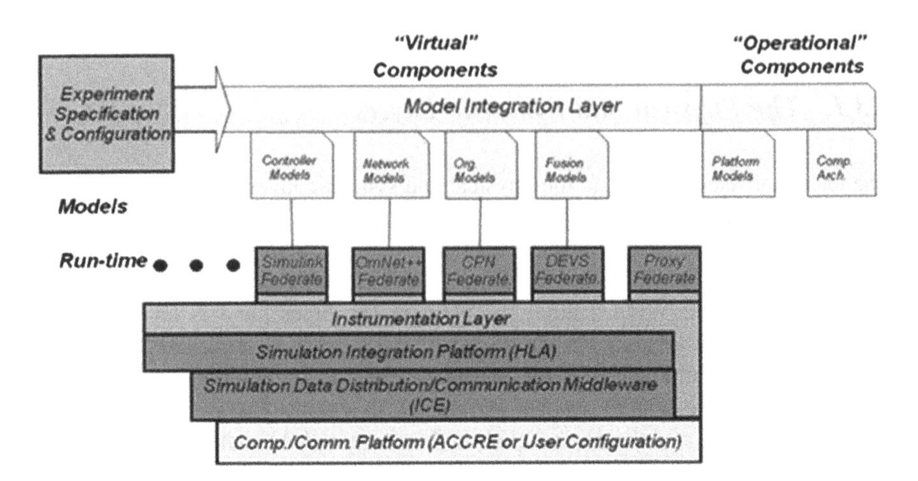

Fig. 5.5 The C2WT architecture

experiments on the HLA platform deployed on a distributed computing environment. After finalizing the component models, the integration models and setting the parameters, the MIL model interpreters generate all the necessary configuration information and run-time code. Each modelling language is depicted as a *federate* on which models built using that language run.

Time Management: It is critical to preserve causality with simulations operating at different timescales. The C2WT builds upon the time management features of the underlying HLA standard, which has provision for both discrete time and discrete event models. The main elements of time management in HLA are: (a) a Logical Timeline, (b) Time ordered delivery of interactions between simulations and (c) a protocol for advancing Logical Time. In a causality preserving execution (note that HLA supports untimed executions as well), the underlying RTI maintains a logical time, and interaction messages generated by simulations are time stamped with the logical time and delivered to their destinations in a timed order. The logical time is advanced by a cooperative Time Advance Request and Grant protocol. A similar protocol is supported for event-driven simulations in which the event-driven simulation requests the Next Event to the RTI. The simulation logical time is advanced either to the earliest available interaction or to the time stamp of the next event local to the requesting simulation.

One can also envision a federation of two or more of these test bed infrastructures through the incorporation of the different modelling languages as federates. However, while this is sufficient to enable the passing of data from one model to another and for the computational and syntactical interoperation of models, it is not sufficient to ensure the semantic and mathematical/algorithmic interoperability. For that, basic research was needed on meta-modelling.

5.3.2 The Semantic Layer and Meta-Modelling

A meta-model is an abstraction layer above the actual model and describes the modelling formalism used to create the model. Consequently, a model has to conform to its meta-model in the way that a computer program conforms to the grammar of the programming language in which it is written. Meta-modelling is defined to be the process of constructing a meta-model in order to model a specific problem within a certain domain. The typical role of a meta-model is to define how model elements are instantiated.

In an effort to study and formally represent the semantic interoperability of disparate models and modelling languages, Rafi [26] and Levis et al. [19] have developed a meta-modelling framework. This meta-modelling approach extends earlier works by Kappel et al. [14] and Saeki and Kaiya [27] for a class of modelling languages primarily used for behavioural modelling problems.

The meta-modelling approach presented here is based on the analysis of the conceptual foundations of a model ensemble so that individual models constructed to address a specific problem in a domain of interest can be evaluated for possible interoperation. The interoperation may be in the form of possible use of the same input data and/or exchange of parameter values or analysis results across different models. This meta-modelling approach provides a framework for identifying these integration mappings between different modelling formalisms especially when they are employed to construct models addressing a common situation or problem of interest. It is a phased approach that uses concept maps, meta-models and ontologies. It is based on comparing the ontologies (for each modelling technique) to help identify the similarities, overlaps and/or mappings across the model types under consideration. The modelling languages considered thus far are Social Networks, Dynamic Meta-Networks and probabilistic decision models such as Influence and Timed Influence Nets. Additional work has been done for Colored Petri Nets modelling decision-making organizations and discrete-event communication network system models.

The approach starts by specifying a modelling paradigm by constructing a generalized concept map [11, 22] that captures the assumptions, definitions, elements and their properties and relationships relevant to the paradigm. An example of a concept map for a timed influence et (TIN), a variant of Bayes nets, is shown in Fig. 5.6. This concept model is a structured representation, albeit not a formal one, and, therefore, not amenable to machine reasoning.

The concept map representation is then formalized using meta-models. This is shown in Fig. 5.7 where the concept map for TIN has been expressed formally using the unified modelling language (UML). The aim of constructing the meta-model is to reveal the structural (i.e. syntactic) aspects of the modelling formalism and to lay down the foundation for its ontology.

A basic ontology, referred to as a *pseudo ontology*, is constructed which mirrors the meta-model and serves as the foundation ontology; it does not contain any semantic concepts (related to the modelling formalism and to the modelled domain) but acts as the skeleton for the ontology.

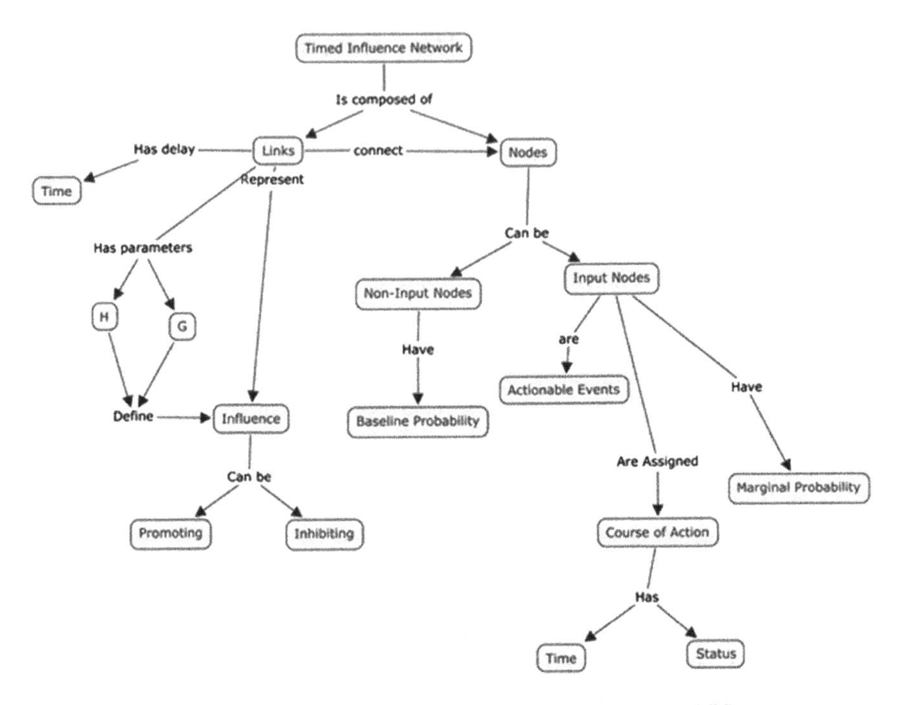

Fig. 5.6 Concept map for timed influence nets (TIN) timed influence net (TIN)

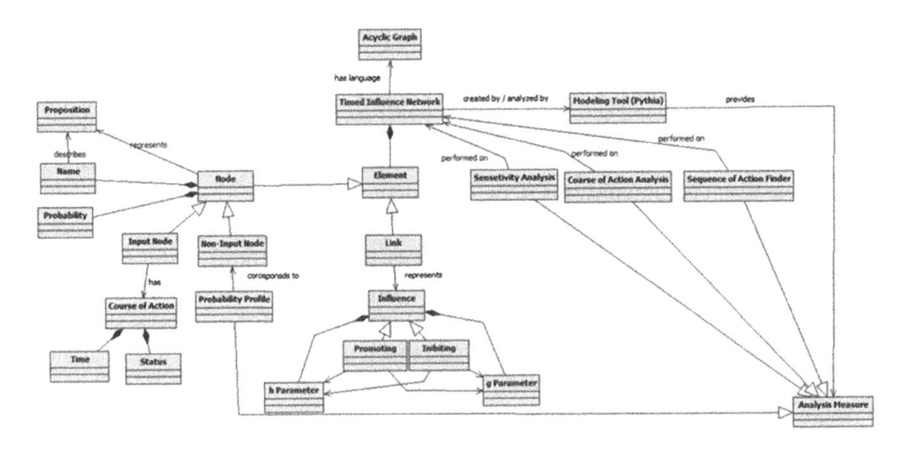

Fig. 5.7 The TIN concept map expressed in UML

In the next step, semantic concepts and relationships are added to this foundation ontology to obtain the *refactored ontology*. Once the individual ontologies are completed for each modelling technique, mapping of concepts across the ontologies is started. The resulting ontology which contains these concepts and relationships within and across multiple ontologies is called an *enriched ontology*.

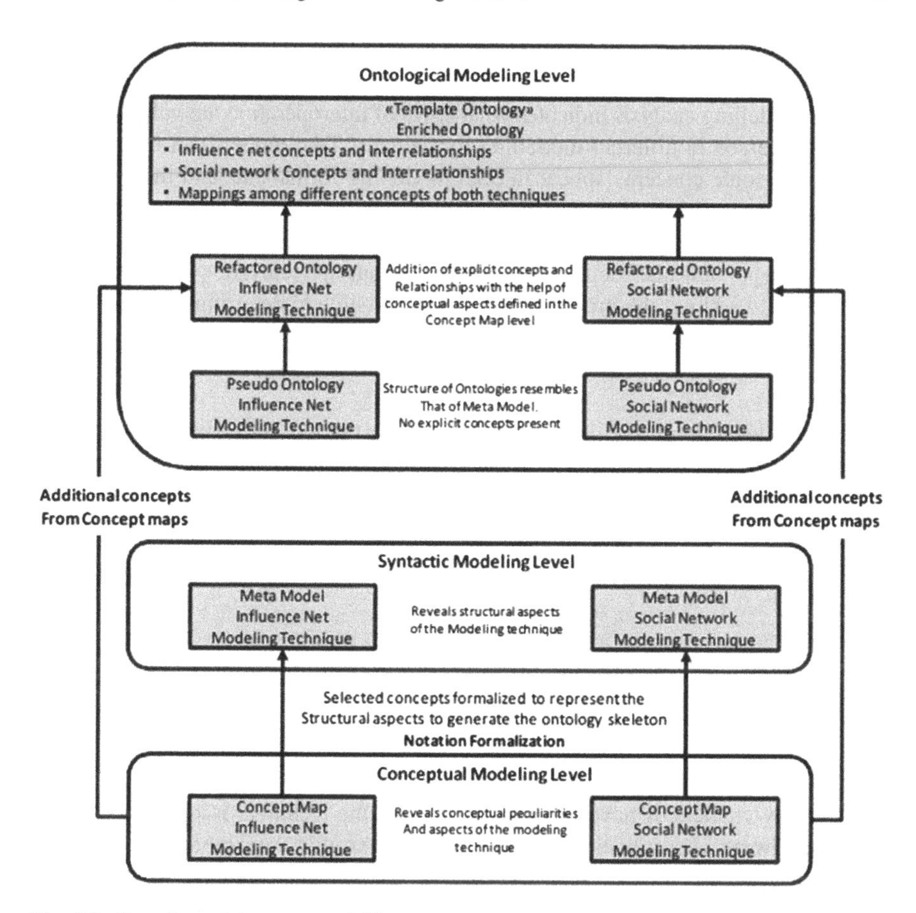

Fig. 5.8 Overview of the meta-modelling approach

Figure 5.8 provides an overview of the workflow prescribed by the meta-modelling approach when applied to two modelling formalisms: Timed Influence Nets and Social Networks. The enriched ontology so constructed for the modelling formalisms can be reasoned with using the logical theory supporting the ontological representation [2]. This mapping suggests ways in which several modelling formalisms can interoperate on a multi-model integration platform such as the C2WT. The mappings suggest possible semantically correct ways to ensure consistency and to exchange information (i.e. parameter values and/or analysis results) between different types of models when they are used in a workflow solving a specific problem of interest.

Analysis of the network structure of the concept map can provide guidance as to the conditions under which the concept map is sufficiently complete to assess model interoperability. For example, social network metrics may be used; Concepts Maps can be imported in a tool such as ORA for assessment [5]. Furthermore, by comparing the network structure of concept maps for different modelling formalisms it may be possible to identify previously unidentified connections among the resulting models.

Meta-modelling analysis indicates what types of interoperation are valid between models expressed in different modelling formalisms. Two models can interoperate (partially) if some concepts appear in both modelling formalisms and they have no contradictory concepts that are invoked by the particular application. By refining this approach to partition the concepts into modelling formalism/language input and output concepts and also defining the concepts that are relevant to the questions being asked to address the problem, it becomes possible to determine which sets of models can interoperate to address some or all of the concepts of interest, and which sets of models use different input and output concepts that are relevant to those questions.

In order to support *semantic interoperability,* models will need to be interchanged across tools. This requires model transformations. The transformations are formally specified in terms of the meta-models of the inputs and the outputs of the transformations. From these meta-models and the specification of the semantic mapping, a semantic translator that implements the model transformation is synthesized [6].

5.3.3 The Workflow Layer

In achieving multi-formalism modelling (MFM), and to provide supporting platforms, many challenges have to be faced. Beside the technical issues that usually arise in allowing interoperations between models through their modelling tools, as described in the previous subsections, there is also a major challenge of improving the human interface to the MFM process itself [7].

A systematic methodology for developing and then using a Domain-Specific MFM Workflow Language (DSMWL) is needed. The objective is to help users of MFM platforms in creating workflows of modelling activities while guaranteeing both syntactic and semantic correctness of the resulting ensemble of interoperating models. The approach is domain specific; the rationale behind this is twofold: first, problems to be solved by employing MFM techniques are usually domain specific themselves; second, it narrows down the scope of meaningful interoperations among several modelling formalisms where each formalism offers unique insights and make specific assumptions about the domain being modelled (see discussion of Fig. 5.3). The first step consists of the identification and characterization of a domain of interest and the modelling techniques that support it. This defines a region in containing a number of nonempty cells in the construct of Fig. 5.3. Domain Analysis follows; its aim is to provide formal representations of syntactic and semantic aspects of the domain. A new Domain-Specific MFM Workflow Language is then developed to construct workflows that capture MFM activities in the selected domain. A domain Ontology resulting from the Domain Analysis step is utilized to provide semantic guidance that effects valid model interoperation.

Domain-Specific Modelling Languages (DSMLs) are languages tailored to a specific domain. They offer a high level of expressiveness and ease of use compared with a General Purpose Language (GPL). Development of a new DSML is not a straightforward activity. It requires both domain knowledge and language development expertise. In the first place, DSMLs were developed simply because they can offer domain-specificity in better ways. According to Mernick et al. [21], the development of any new DSML should go in five phases: Decision, Analysis, Design, Implementation and Deployment. The details of the process for developing a DSMWL are documented in Abu Jbara [1] and Levis and Abu Jbara [17].

Creating workflows using a domain-specific language allows for translating visual views of model interoperation into an executable implementation. There already exist generic techniques for creating and executing workflows such as BPMN [23] and BPEL [24]. The domain-specific nature of the MFM approach requires the development of a Domain-specific MFM Workflow Language for the selected domain of interest. Such a language would be tailored to a problem domain of interest and would offer a high level of expressiveness. It can be a specific profile of an existing General Purpose Language, i.e. BPMN.

5.3.4 The Approach

For effective use of multi-formalism modelling, all four layers are necessary. Implementation of the physical and syntactic layers results in a computational testbed; the semantic layer through metamodelling analysis addresses the validity of model interoperation; and the workflow layer, through the Domain-Specific multi-formalism modelling Workflow Languages it contains, enables the construction and management of a particular multi-formalism construct to address the problem of interest. Based on this foundation, it is possible now to consider a number of different modelling formalisms that can be used to explore questions involving human organizations.

5.4 Modelling Human Organizations

A suite of modelling tools is shown in Fig. 5.9, each tool conforming to a different modelling formalism. They have been developed by the System Architectures Laboratory (SAL) of George Mason University and the Center for Computational Analysis of Social and Organizational Systems (CASOS) of Carnegie Mellon University.

For example, social networks [4] describe the interactions (and linkages) among group members but say little about the underlying organization and/or command structure. **ORA** [5] is an application for the construction and analysis of social

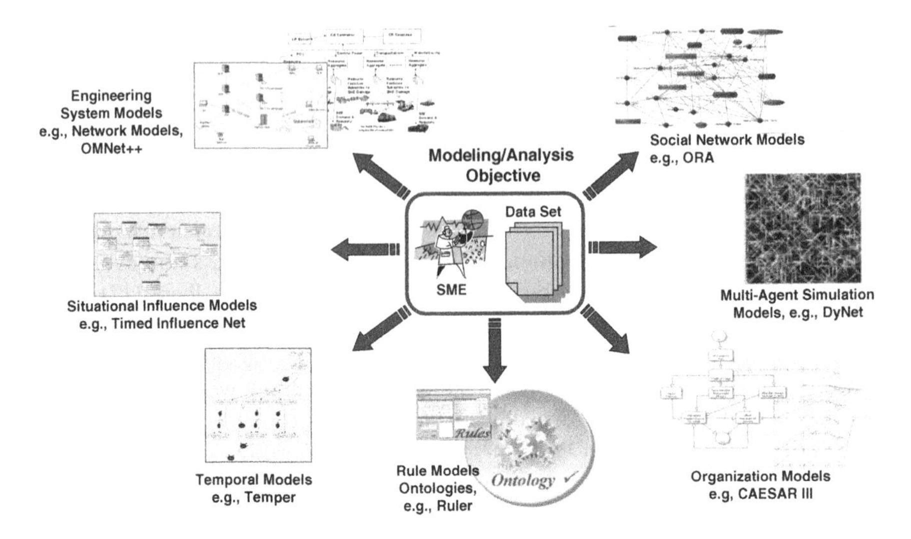

Fig. 5.9 Tools for modelling human organizations

networks, while ***DyNet*** [3] is a multi-agent simulation model for studying dynamic changes in human networks.

Similarly, organization models [15] focus on the structure of the organization and the prescribed interactions but say little on the social/behavioural aspects of the members of the organization. ***Caesar III*** is a Colored Petri Net tool for designing and analyzing organizational structures [18]. Policies and procedures that govern the behaviour of an organization need to be modelled and analyzed because they affect the behaviour of the human organization. ***Ruler*** is a tool for evaluating whether a proposed course of action is in compliance with the prevailing policies and procedures [30]. ***Temper*** is a temporal logic inference tool that is used to address the temporal aspects of a course of action [29].

Timed Influence Net (TIN) models [28] are a variant of Bayesian models, used to describe cause and effect relationships among groups at a high level. The Timed Influence Net application, ***Pythia*** [16], is used to develop courses of action and compare their outcomes.

In addition to these tools, existing tools can be incorporated such as a tool for modelling communications networks (e.g. the open source OMNeT++ [25] and the INET Framework [12]).

5.5 Interdiction of Drug Traffickers

In this section, an application of the Multi-Formalism Modelling approach to problems involving human organizations is described. It is based on the operations of an actual organization, the US Joint Interagency Task Force—South, whose mission is interdiction of drug trafficking in the southeastern part of the US. The scenario was inspired by a motion picture: *Contraband* (2012). It uses three modelling formalisms: Social Networks with **ORA**; Colored Petri Nets with **CAESAR III** and Timed Influence Nets with ***Pythia*** as well as a Geographic Information System (**GIS**) for visualization.

JIATF-South is a Drug Interdiction agency well known for interagency cooperation and intelligence fusion. The agency usually receives disparate data regarding drug trafficking from different sources. Quick and effective analysis of data is essential in addressing drug trafficking threats effectively. A typical case begins with JIATF-South receiving information form the US Drug Enforcement Administration. This prompts the deployment of drones that subsequently detect and monitor a suspect vessel until JIATF-South can sortie a Coast Guard cutter to intercept. If drugs are found, jurisdiction and disposition over the vessel, drugs and crew are coordinated with other national and international agencies. Courses of Action (COAs) identified by JIATF-South are fully dependent on efficient analysis of received data.

Analysts at JIATF-South are trained to use various modelling techniques to analyze data and then to identify possible COAs. By applying the MFM approach, analysts would be capable of creating visual workflows supported by interoperating models. An informal description of the domain is as follows:

- Drug Interdiction involves information sharing, fusion of intelligence data and monitoring of drug trafficking activities.
- Drug Interdiction involves dealing with Drug Cartels and Smugglers and Law Enforcement and Intelligence.
- Drug Smugglers takes different routes and originate from different geographical points.
- Analysts use Social Networks, Organization Models, Influence Nets, GIS and Asset Allocation and Scheduling techniques.
- Given incomplete and uncertain information, timely decisions need to be made on best Courses of Action.

After identifying related concepts, a Concept Map is constructed to capture the relations between the concepts. Figure 5.10 shows a concept map that addresses the question: How does JIATF-South perform Drug Interdiction?

Domain analysis follows by using the generated concept maps to construct UML class diagrams that represent the constructs of the domain. Using GME, a meta-model for this Domain's workflow MFM language is defined based on the UML class diagram resulting from the domain analysis. This meta-Model defines the constructs of this new language. Figure 5.11 shows part of the visualization aspect of the Domain-specific MFM Workflow Language. In addition to the basic constructs borrowed from BPMN, some new constructs have been introduced and some constraints have been imposed. The resulting workflow has two types of activities, operations and interoperations. Operations are those activates performed on a specific model using the modeling tool that supports its modeling formalism. Interoperations are those activities that involve interoperations across models

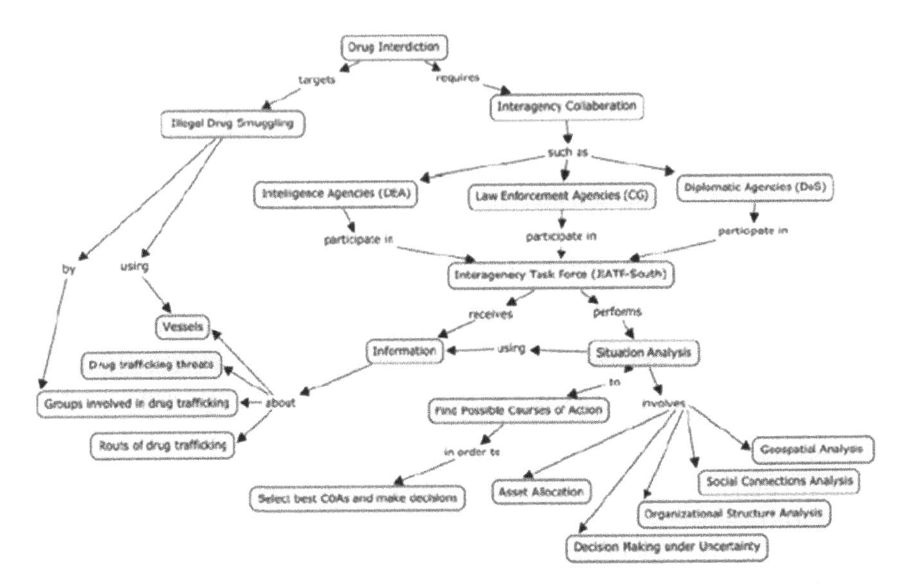

Fig. 5.10 Concept map for drug interdiction

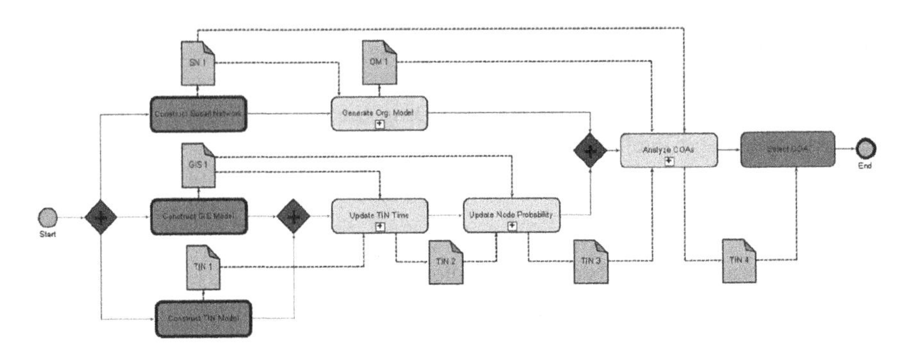

Fig. 5.11 Workflow of drug interdiction MFM activity expressed in the domain-specific MFM workflow language

through their modelling tools. Operations in this DSMWL can be in one of two flavours, Thick or Thin. This is due to the fact that multi-modelling platforms can support the integration of modelling tools in one of two forms. Thin operations represent the case when service based integration takes place, given that the modelling tool of interest exposes its functionalities as services. Thick operations represent the case in which the whole modelling tool is integrated as a package in the multi-modelling platform.

Once the GME meta-model of the constructed Domain Specific Multi-Modelling Workflow Language is interpreted and registered as a new modelling paradigm in GME, the GME environment is used to create workflows that capture specific domain scenarios. Figure 5.11 shows a workflow that involves the use of Geospatial models, Timed Influence Nets, Social Networks and Organization Models to analyze data and then generate and select best Courses of Actions (COAs) for drug interdiction.

The Fictitious Scenario: The JIATF-South agency receives information about suspicious activities from different sources including local, regional and international intelligence agencies. The scenario events begin when a cargo ship with R flag (R is a country in the Caribbean) is being loaded with drugs.

A drug cartel operating in country R is responsible for this drug smuggling activity. The local intelligence agency of country R intercepts a phone call between a person known to be the head of the cartel in country R and a customs officer in R's port authority. R's intelligence agency shares information with JIATF-S; its officers react directly and begin analyzing the information. The suspected drug cartel in country R is already known to the officers. It is also known that this R-based cartel is connected to a US-based cartel. JIATF-S has an insider informant in the US-based cartel; the informant is requested to provide more information about this particular case.

The cargo ship leaves the port of country R on Day n and enters international waters on day n + 1. JIATF-S has drones that continuously monitor suspicious activities. Orders go out from JIATF-S directing some drones to monitor the suspected cargo ship and to keep it under surveillance. The cargo ship is expected to arrive to a US port on the Gulf of Mexico on day n + 5. A visualization of the scenario is shown in Fig. 5.12.

Based on the information available, JIATF-S analysts construct a TIN model in ***Pythia*** and use it to analyze the effects of actions that can be taken to address the drug trafficking threat in this scenario. Figure 5.13 shows the model created and analyzed as part of the scenario workflow. The goal, as shown on the node on the right side of the model, is to interdict drug trafficking activity. The actions that can be taken by JIATF-S and other scenario actors are shown on the left side of the model. Pythia has the capability to analyze the effect of setting each of these actions on or off and determining the optimum time for applying these actions. TIN analysis results in a set of probability profiles for each of the effects of interest when a particular Course of Action is applied. The COA that maximizes the probability of the goal node, interdicting drug trafficking, is considered to be the best COA. This was obtained using the SAF optimization algorithm embedded in ***Pythia***. The selected COA identifies which actionable events influence the success of drug trafficking efforts.

A Social Network model captures relationships between agents. JIATF-S receives continuous information about drug cartels and their members. With the help of **ORA**, relationships between cartel's members and among different cartels are captured (Fig. 5.14). These networks can help to identify the most effective individuals, "leaders" and the paths of communication between different cartels. Figure 5.14 shows a simple Social Networks for the cartels involved in the scenario. The nodes on the left side of the model represent the cartel based on country R, while the nodes on the right represent the US-based cartel. The arrows represent

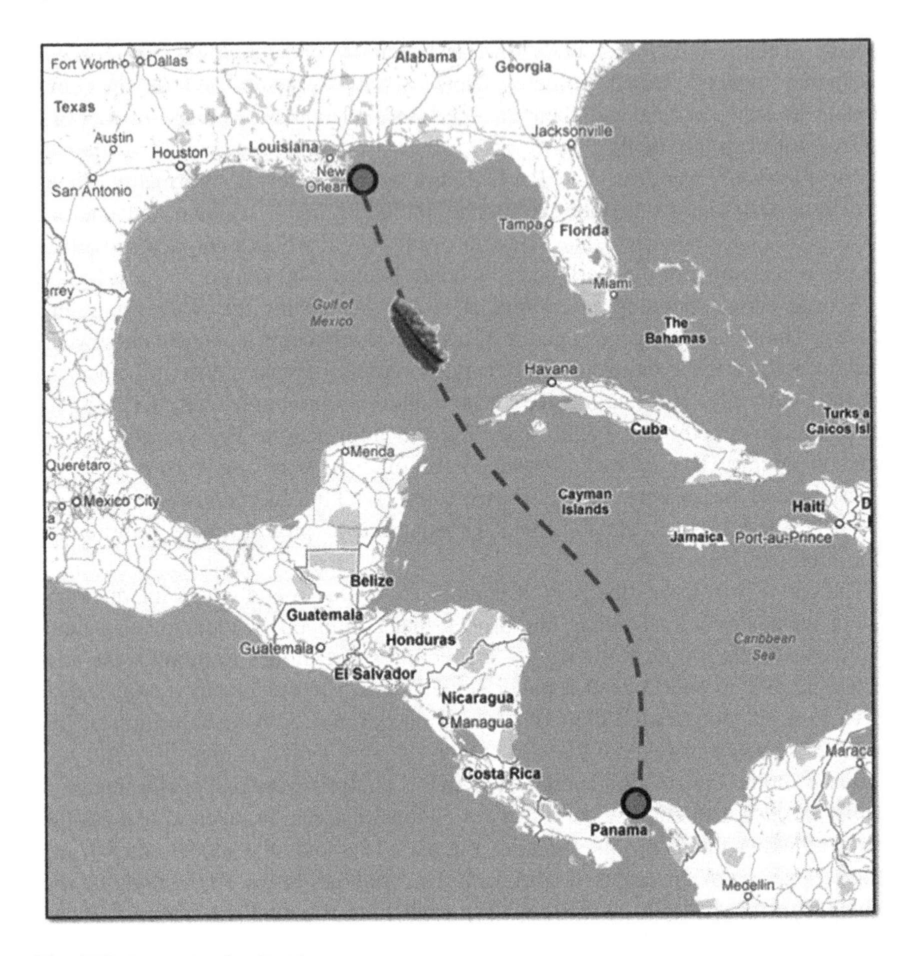

Fig. 5.12 Scenario visualization

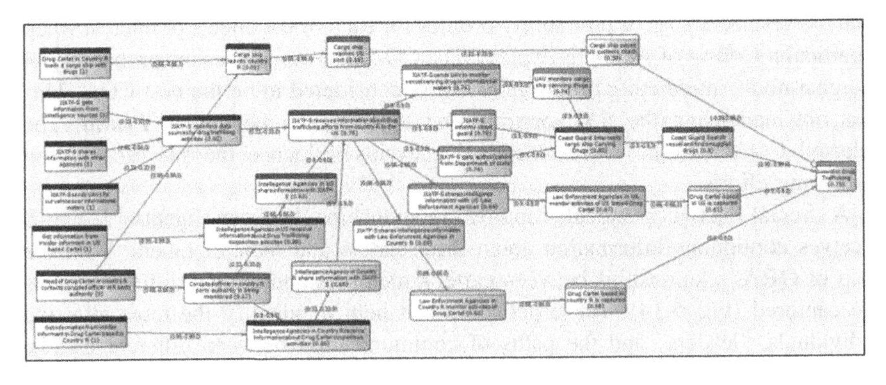

Fig. 5.13 Timed influence net model for developing drug interdiction courses of action

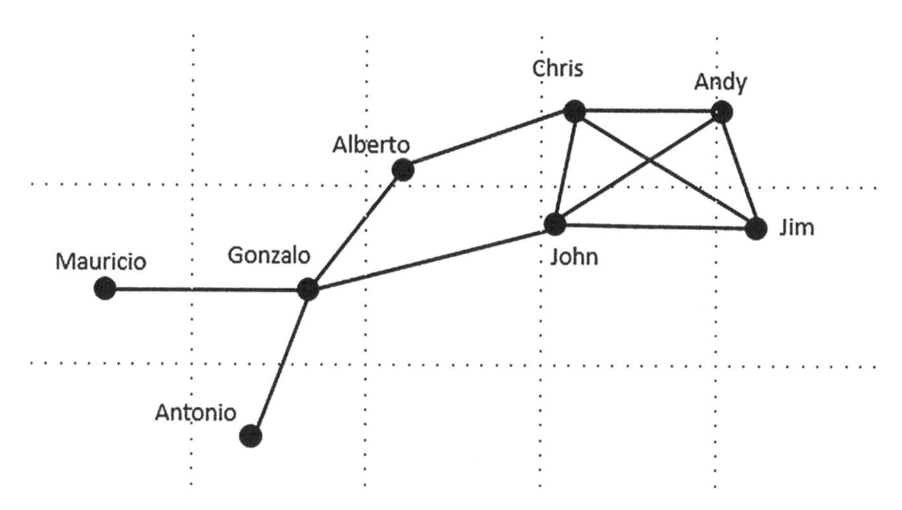

Fig. 5.14 Social network of country R and US drug cartels

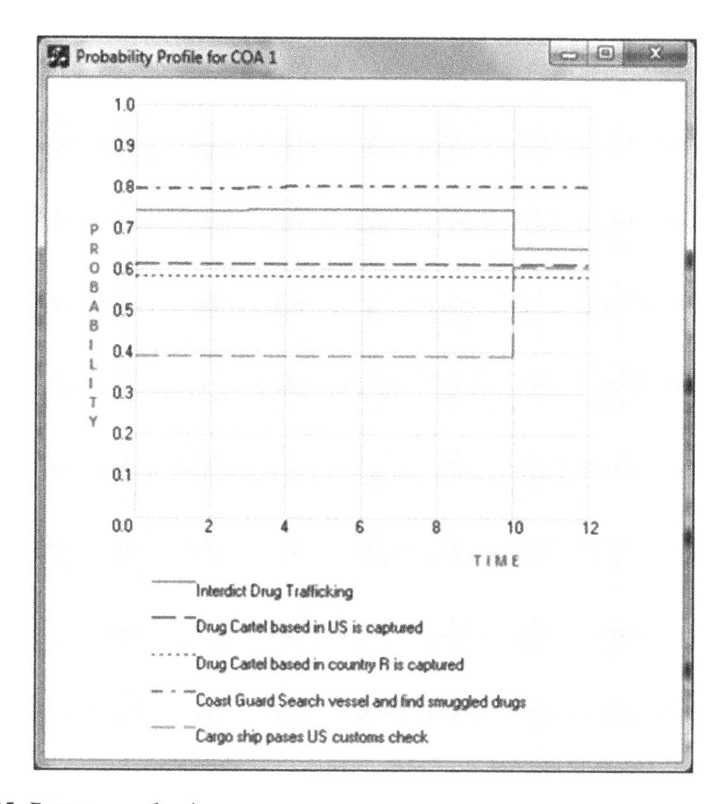

Fig. 5.15 Best course of action

the relationships between the individuals and help to identify the most connected and influential individuals. These kinds of relationships and information captured in social networks help JIATF-S analysts to identify best actions to track and interdict drug trafficking activities taken by these two cartels. In the scenario workflow, a social network is used to generate an organization structure that is modelled as a Colored Petri Net using **CAESAR III** for further analysis and then to revise the TIN model before using it to generate COAs.

After the workflow of the interoperating models is interpreted, it gets executed on the testbed. The participating modelling techniques are utilized through their supporting tools that are already integrated into the computational platform. The focal point of the models is the TIN model. The workflow sequence operates on revising and optimizing this TIN model for better COA analysis and selection. Interoperations between models participating into the multi-modelling activity enhance the analysis process. The relationships captured in the social network model identify the most influential actors in the drug trafficking effort and are used to update the TIN model actions and probabilities. Temporal information captured in the geospatial model is used to update the timing of actions in the TIN model.

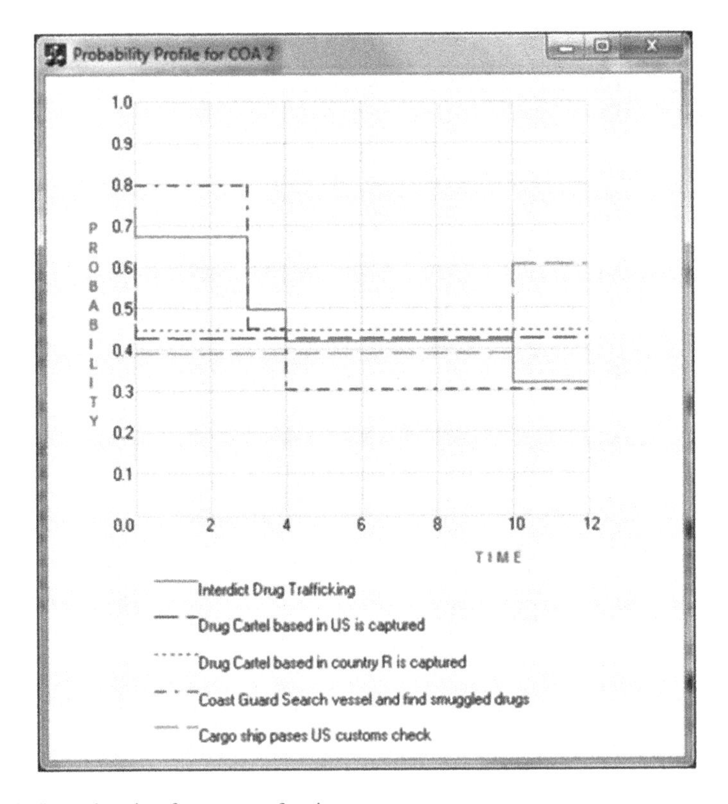

Fig. 5.16 Second option for course of action

After the TIN model is refined based on the data received from the other models, ***Pythia*** capabilities and algorithms are utilized for COA selection.

The last operation of the workflow, as was shown in Fig. 5.11, represents the COA selection task. Different combinations of actions are examined to determine the COA that gives highest probability for a specific node goal, which is successful drug interdiction in this scenario. Three generated COAs are presented to illustrate the types of results obtained.

The first COA, visualized in Fig. 5.15, shows that sharing information between the JIATF-South and other (local and regional) intelligence agencies in addition to the utilization of surveillance resources, results in the probability of the target node of effective drug interdiction reaching its highest level around 68 %. This is the selected COA by JIATF-S since it maximizes the probability of Drug Interdiction for the scenario under consideration.

In the second COA (Fig. 5.16), the probability of interdicting smuggled drugs decreases dramatically to about 32 % when information sharing between the JIATF-South and other local and regional intelligence agencies is not in effect. *This shows the value of information sharing to the success of drug interdiction efforts.*

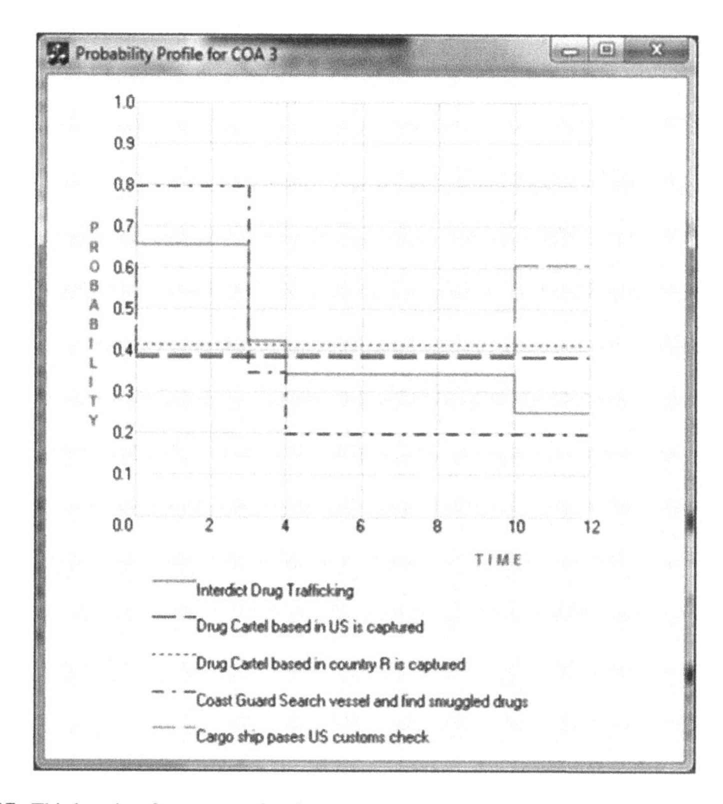

Fig. 5.17 Third option for course of action

The third COA (Fig. 5.17) shows how the probability of effective drug interdiction can decrease even more to a level close to 25 % if in addition to lack of information sharing, insider information from the drug cartel is not available.

5.6 Conclusion

A methodology for employing multi-formalism modelling effectively has been described. The four layers of the approach have been identified and the issues that need to be considered at each layer have been described. The approach has been used in a number of diverse applications where a variety of different modelling formalisms have been employed. One particular application focused on the interactions of human organizations is included to illustrate the process.

Acknowledgement The contribution of Prof. Kathleen Carley and her students, Carnegie Mellon University, in the development of some of the underlying ideas while working together in collaborative projects is gratefully acknowledged. This chapter is based on the work of many students and colleagues in the System Architectures Laboratory at George Mason University, especially Drs. Zaidi, Wagenhals and Abu Jbara.

References

1. Abu Jbara A (2013) On using multi-modeling and meta-modeling to address complex systems. Ph.D. Dissertation, George Mason University, Fairfax, VA
2. Bechhofer S (2003) OWL reasoning examples, University of Manchester, http://owl.man.ac.uk/2003/why/latest/
3. Belov N, Martin MK, Patti J, Reminga J, Pawlowski A, Carley KM (2009) Dynamic networks: rapid assessment of changing scenarios. In: Proceedings of the 2nd international workshop on social computing, behavior modeling, and prediction, Phoenix, AZ, Apr
4. Carley KM (1999) On the evolution of social and organizational networks. In: Andrews SB, Knoke D (eds) Special issue of Research in the sociology of organizations, vol 16, pp 3–30
5. Carley KM, Pfeffer J, Reminga J, Storrick J, Columbus D (2013) ORA user's guide 2013, Carnegie Mellon University, School of Computer Science, Institute for Software Research, Technical Report, CMU-ISR-13-108
6. Emerson M, Sztipanovits J (2006) Techniques for metamodel composition. In: Proceedings of the 6th OOPSLA workshop on domain-specific modeling, Portland, OR, Oct 22, pp 123–139
7. Fishwick PA (2004) Toward an integrative multimodeling interface: a human-computer interface approach to interrelating model structures. SCS Trans Model Simul 80(9):421–32
8. Garlan D, Carley KM, Schmerl B, Bigrigg M, Celiku O (2009) Using service-oriented architectures for socio-cultural analysis. In: Proceedings of the 21st international conference on software engineering and knowledge engineering (SEKE2009), Boston, MA, July 1–3
9. Haider S, Levis AH (2007) Effective course-of-action determination to achieve desired effects. IEEE Trans Syst Man Cybern Part A: Syst Hum 37(6):1140–1150
10. Hemingway G, Neema H, Nine H, Sztipanovits J, Karsai G (2011) Rapid synthesis of high-level architecture-based heterogeneous simulation: a model-based integration approach. Simulation 17 March
11. IHMC (2014) IHMC CMap tools http://cmap.ihmc.us/download/index.php

12. INET Framework (2014) www.inet.omnetpp.org
13. Karsai G, Maroti M, Lédeczi A, Gray J, Sztipanovits J (2004) Composition and cloning in modeling and meta-modeling. IEEE Trans Control Syst Technol 12(2):263–278
14. Kappel G, Kapsammer E, Kargl H, Kramler G, Reiter T, Retschitzegger W, Schwinger W, Wimmer M (2006) Lifting metamodels to ontologies: a step to the semantic integration of modeling languages. MoDELS 528–542
15. Levis AH (2005) Executable models of decision making organizations. In: Rouse WB, Boff K (eds) Organizational simulation. Wiley, New York
16. Levis AH (2014) Pythia 1.8 user's manual. System Architectures Laboratory, George Mason University, Fairfax, VA
17. Levis AH, Abu Jbara A (2013) Multi-modeling, meta-modeling and workflow languages. In: Gribaudo M, Iacono M (eds) Theory and application of multi-formalism modeling. IGI Global, Hershey, PA
18. Levis AH, Kansal SK, Olmez AE, AbuSharekh AM (2008) Computational models of multinational organizations. In: Liu H, Salerno JL, Young MJ (eds) Social computing, behavioral modeling and prediction. Springer, New York
19. Levis AH, Zaidi AK, Rafi MF (2012) Multi-modeling and meta-modeling of human organizations. In: Salvendy G, Karwowski W (eds) Advances in human factors and ergonomics 2012: proceedings of the 4th AHFE conference, San Francisco, CA. CRC Press, Boca Raton, FL, pp 21–25
20. Maria A (1997) Introduction to modeling and simulation. In: Proceedings of the 29th Winter simulation conference, Atlanta, GA, Dec 7–10. IEEE Computer Society, Washington, D.C
21. Mernik M, Heering J, Sloane AM (2005) When and how to develop domain-specific languages. ACM Comput Surv (CSUR) 37:316–344
22. Novak JD, Cañas AJ (2008) The theory underlying concept maps and how to construct and use them. Technical Report IHMC CmapTools 2006-01 Rev 01-2008, Pensacola, FL
23. OMG-BPMN (2011) Business Process Model and Notation (BPMN). http://www.omg.org/spec/BPMN/
24. OMG-BPEL (2014) Business Process Execution Language (BPEL). https://www.oasis-open.org/committees/wsbpel/
25. OMNeT ++ (2014) www.omnetpp.org
26. Rafi MF (2010) Meta-modeling for multi-model integration. MS Thesis, Department of Computer Science, George Mason University, Fairfax, VA (May)
27. Saeki M, Kaiya H (2006) On relationships among models, meta models and ontologies. In: Proceedings of the 6th OOPSLA workshop on domain-specific modeling, Portland, OR, Oct 22
28. Wagenhals LW, Levis AH (2007) Course of action analysis in a cultural landscape using influence nets. In: Proceedings IEEE symposium on computational intelligence in security and defense applications, CISDA, Honolulu, Hawaii, Apr 1–5, pp 116–123
29. Zaidi AK, Levis AH (2001) TEMPER: a temporal programmer for time-sensitive control of discrete event systems. IEEE Trans Syst Man Cybern 31(6):485–96
30. Zaidi AK, Levis AH (1997) Validation and verification of decision making rules. Automatica 33(2):155–169

Author Biography

Alexander H. Levis is University Professor of Electrical, Computer and Systems Engineering and heads the System Architectures Laboratory at George Mason University, Fairfax, VA. From 2001 to 2004, he served as the Chief Scientist of the U.S. Air Force. He was educated at Ripon College where he received the AB degree (1963) in Mathematics and Physics and then at MIT where he received the BS (1963), MS (1965), ME (1967) and Sc.D. (1968) degrees in Mechanical Engineering with control systems as his area of specialization. For the past 15 years, his areas of research have been architecture design and evaluation, resilient architectures for command and control and multi-modelling for behavioural analysis. Dr. Levis is a Life Fellow of the Institute of Electrical and Electronic Engineers (IEEE) and past president of the IEEE Control Systems Society; a Fellow of the American Association for the Advancement of Science and of the International Council on Systems Engineering and an Associate Fellow of the American Institute of Aeronautics and Astronautics. He has over 280 publications documenting his research, including the three volume set that he co-edited on *The Science of Command and Control*, and *The Limitless Sky: Air Force Science and Technology contributions to the Nation* published in 2004 by the Air Force.

Chapter 6
A New Research Architecture for the Simulation Era

Martin Ihrig

Abstract This chapter proposes a novel research architecture for social scientists who want to employ simulation methods. The new framework gives an integrated view of a research process that involves simulation modelling. It highlights the importance of the theoretical foundation of a simulation model and shows how new theory-driven hypotheses can be derived that are empirically testable. The paper describes the different aspects of the framework in detail and shows how it can help structure the research efforts of scholars interested in using simulations.

Keywords Simulation · Modelling · Research · Methodology · Business · Management · Theory · Social sciences

6.1 Introduction

Business and management researchers are increasingly interested in exploring phenomena that are emergent and/or one of a kind and in studying complex and non-repeatable processes. Simulation modelling is the appropriate methodological approach for this kind of research. Harrison et al. [7, p. 1229] consider simulation modelling to be a "powerful methodology for advancing theory and research on complex behaviors and systems", while Davis et al. [4, 480] point out that "the primary value of simulation occurs in creative and systematic experimentation to produce novel theory". Simulation research results in theory-driven frameworks and hypotheses that would be difficult to obtain from empirical analyses alone. In this chapter, we propose a novel research architecture, a framework that gives an integrated view of a research process that involves simulation modelling. In what follows, we describe the different aspects of the framework and show how it can help structure the research efforts of scholars interested in using simulations.

M. Ihrig (✉)
University of Pennsylvania, 418 Vance Hall, 3733 Spruce Street, Philadelphia, PA 19104, USA
e-mail: ihrig@wharton.upenn.edu

K. Al-Begain and A. Bargiela (eds.), *Seminal Contributions to Modelling and Simulation*, Simulation Foundations, Methods and Applications,
DOI 10.1007/978-3-319-33786-9_6

6.2 An 'Extended' Logic of Simulation as a Method

Carley [1, 254] gives a detailed explanation for "why so many social and organizational scientists and practitioners are turning to computational modeling and analysis as a way of developing theory and addressing policy issues". Among the many reasons she advances is that "social and organizational systems are complex non-linear dynamic systems; hence, computational analysis is an appropriate technology as models can have these same features." [1, 254] Harrison, Lin, Carroll and Carley note that "the academic field of management has been slow to take advantage of simulation methods" [7, p. 1229]. But looking at the increased number of recent articles based on simulation research in management journals and of simulation-specific workshops and papers at management conferences, it seems that management theorists are finally discovering the benefits of simulation methods. Davis et al. [4] and Harrison et al. [7] give guidelines for simulation research in the field of management. Gilbert and Troitzsch [6] put forward the following framework (Fig. 6.1) to explain the logic of simulation as a method in their authoritative book on simulation for the social scientist.

Starting at the bottom left, the real-world 'target' under study is modelled by abstracting characteristics from it, and then a computer model is used to run simulations in order to produce simulated data. This data can then be compared with data collected in the 'real' social world.

This framework nicely illustrates the core logic of simulation as a method that underlies all the different kinds of simulation approaches that Gilbert and Troitzsch [6] review in their book. But does it capture the entire simulation research process that scholars encounter, especially for complex agent-based modelling approaches? What is the role of existing theory, insights and frameworks from the literature, when it comes to modelling? Where do simulation environments, software toolkits that help researchers create, run and analyze simulation models, come into play? To account for those questions and provide a more detailed picture of the simulation research life cycle, we constructed an expanded framework (Fig. 6.2) to extend the logic of Gilbert and Troitzsch's [6] basic model. This framework is meant to provide future simulation researchers with a research architecture that can assist them in their modelling efforts. Simulation methods are used for studying complex processes; simulation research, being a complex process itself, should therefore be

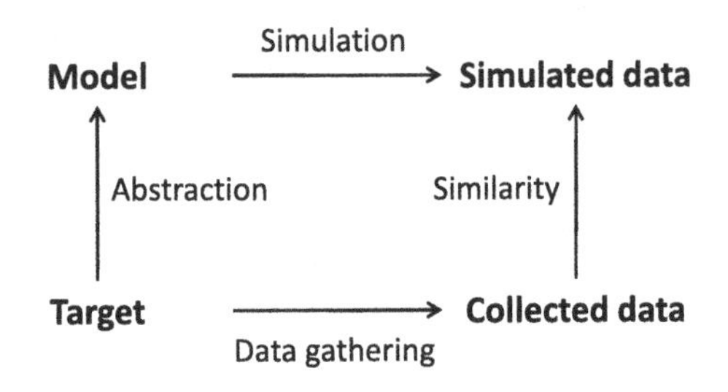

Fig. 6.1 The logic of simulation as a method [6, 17]

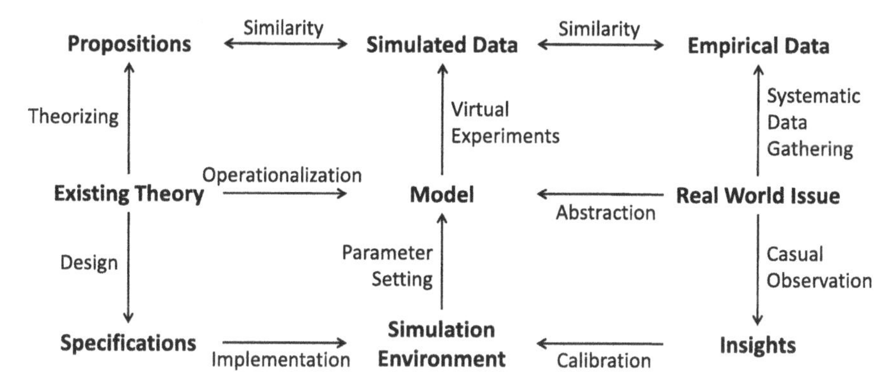

Fig. 6.2 A new research architecture

described by an appropriately complex process model that depicts all the relationships between its building blocks. The following sections explain the different components of this research architecture.

6.3 From Theory to Simulation Model

In contrast to Gilbert and Troitzsch [6], we propose to start the simulation research 'adventure' in the bottom half of the framework (Fig. 6.3) on the left with a particular *theory*. The bottom left square describes the process of implementing theory with software, whereas the bottom right square depicts the real-world grounding of the model.

Implementing Theory with Software: In a very general definition, a theory is "a statement of what causes what, and why" [2]. In order to build simulation software,

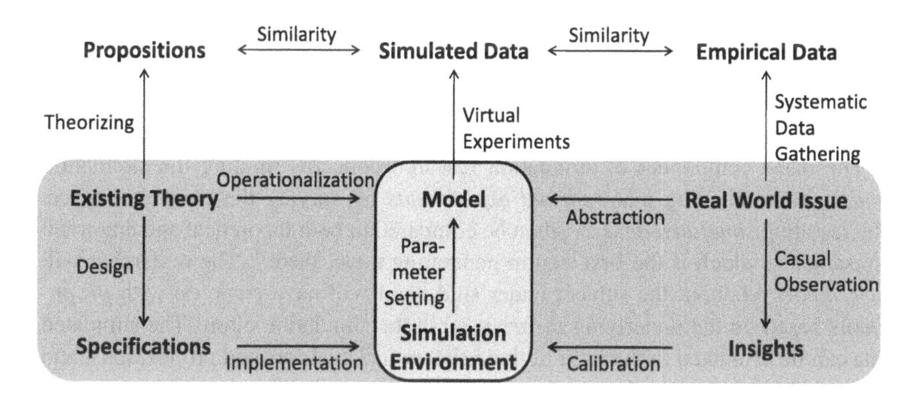

Fig. 6.3 From theory to simulation model

one has to *design* and write a detailed document with *specifications* that mirror the theory in computer algorithms. This is both a conceptual and technical task, and requires the simulation researcher to be very specific about how "what causes what" is based on theoretical considerations ("why"). The next step is to *implement* the concrete specifications in computer code, and program the *simulation environment* as executable software. With the software up and running, finding the right *parameter settings* for the generic simulation environment can allow an application-specific *model* to be created. The resulting model will *operationalize* the specific theory on which the simulation environment is based.

Real-World Grounding: The purpose of simulation research is not necessarily to model a theory in the abstract. The researcher wants to study a *real-world issue* (what Gilbert and Troitzsch [6] call a 'target') so as to produce new theory. *Casual observations* of the social world will yield helpful *insights* about the real-world issue that will help *calibrate* the parameter space in the simulation environment: thus the final model will be an *abstraction* of the real-world issue under study.

What distinguishes the simulation approach described here from many other simulation exercises is that a full-fledged theory lies behind the simulation environment that is built. In the case of agent-based models and simulations [5], most modellers endow their virtual agents with only a couple of simple rules as an abstraction from real social agents (individuals, companies, etc.). Therefore, the entire bottom half of our research framework (and especially the left side) disappears, because the modelling lacks a theoretical underpinning. This is reflected in Gilbert and Troitzsch's [6] simpler framework shown above. More simulation models are needed that have a stronger theoretical foundation, because "the advance of multi-agent techniques provides social scientists, who are used to thinking about agency, the ability to reason in the terms of their theories" [1].

6.4 Simulation Research in Action

Once the researcher has built an application-specific model informed by the real world and grounded in some theory, the actual simulation research can start—the top part of our framework (Fig. 6.4). The top left quadrant describes the process of new theory generation, and the top right links the simulation work to empirical follow-up studies.

The basic component of simulation research concerns running the simulation *model* and conducting many *virtual experiments* by varying the parameter space. The resulting *simulated data* can then be compared to both theoretical and empirical assessments, which is the first step in generating novel theory. The researcher will have *theorized* about the subject under study and will have come up with *propositions* based on the underlying *theory* used in the simulation effort. The simulated data can be evaluated in the light of these theoretical analyses and propositions, and the researcher can learn from studying *similarities* and *differences*. This exercise will result in theory-driven hypotheses that are empirically testable. Subsequently,

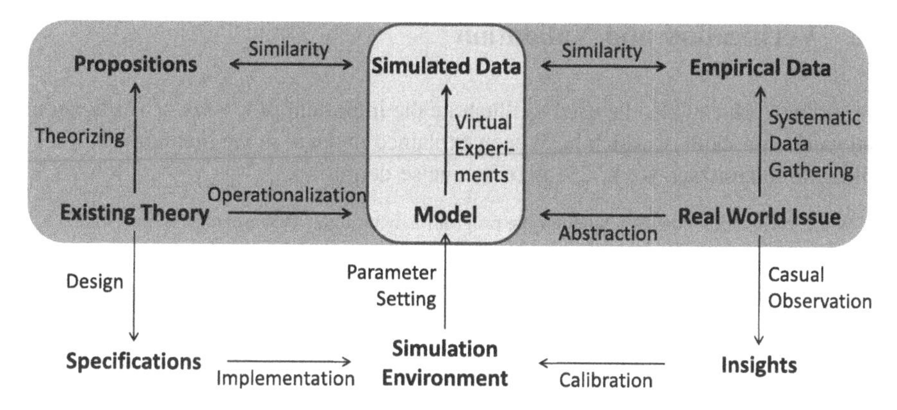

Fig. 6.4 Simulation research in action

the simulated data can be compared to *empirical data*. In an empirical follow-up study, the *real-world issue* that is being investigated can be further examined by obtaining empirical data through *systematic data gathering* on a basis that is informed by the previous simulation research.

6.5 Conventional Research Approach

The research architecture described above is more comprehensive than conventional approaches that do not employ simulation tools, and so is better suited for studying complex phenomena and obtaining new theoretical insights.

The conventional research approach can be depicted by connecting the left- and right-hand ends of our framework (Fig. 6.5). Predictions and analyses are made based on existing theories, and the empirical data gathered on real-world issues is compared to these theoretical accounts or propositions. What is lacking is the power of computer tools that enable us to study more complex processes by modelling micro behaviours that individually might be straightforward, but may result in unpredictable outcomes when considered together.

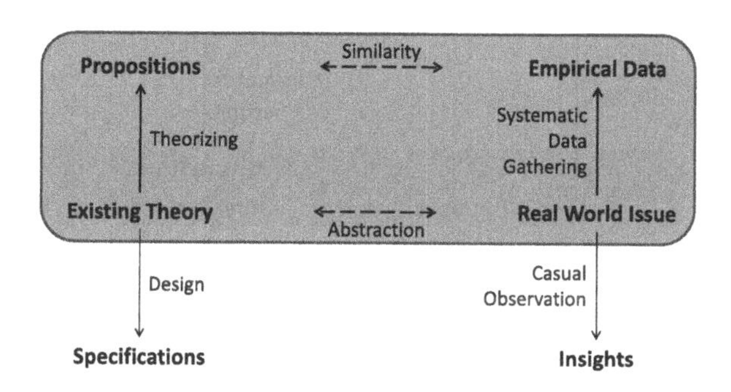

Fig. 6.5 Conventional research approach

6.6 Verification and Validation

Our framework can also be used to illustrate the important processes of verification and validation, both of which have been explained in detail in the literature [1, 4–7]. Gilbert and Troitzsch [6, p. 23] give a concise definition:

> While verification concerns whether the program is working as the researcher expects it to, validation concerns whether the simulation is a good model of the target.

Our framework allows us to identify multiple instances where verification and validation come into play (Fig. 6.6). There are three layers of interest: verification, validation, and ascertaining the two.

Verification. Starting at the bottom left, the obvious area where the model needs verifying is in the building of the software, where the researcher has to ensure that the program's technical specifications have been properly implemented. The simulation environment has to perform exactly as described in the technical document, without errors or 'bugs'. The simulation processes, in their abstracted form, also have to work like and be consistent with the real-world social processes they represent.

Validation: The middle layer of the framework depicts the areas of interest in validation terms. Most researchers will look to the right, and ensure that the application-specific model fully represents the real-world issue or target under study. But there is another important area: since the simulation environment is underpinned by a theory, they also have to make sure the model is a fair representation of the theoretical constructs.

Ascertaining Verification and Validation: The factors noted above will be difficult to ascertain without comparing the simulated data to either empirical data or theoretical predictions, or both. Therefore, simulation researchers have to pay

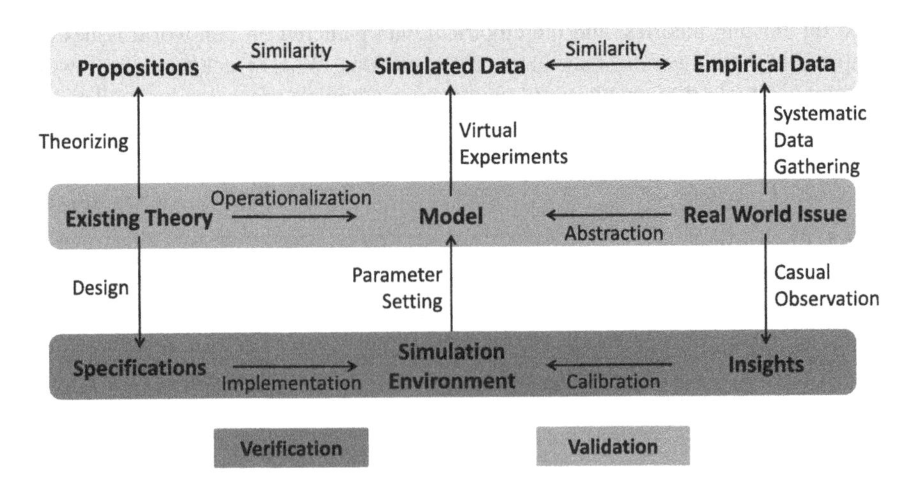

Fig. 6.6 Verification and validation

attention to the top layer of the framework. They have to infer and draw conclusions from the actual results of the simulation runs to assess whether the program is working as intended, and represents the actual phenomenon studied. Great care must be taken here because—as Gilbert and Troitzsch [6] point out—mistakes can occur at any step in the research process.

6.7 Simulation Capabilities Beyond a Single Research Project

Considering our research framework further, the top half of Fig. 6.7 (shaded area, blue) maps the area usually covered by research employing simulation methods. The core research activities classically conducted in scholarly work based on simulation modelling as described by Gilbert and Troitzsch [6] are those depicted in the top right quadrant. Many recent simulation studies also base their modelling on existing theories, represented by the top left quadrant (a fine example of this would be Csaszar and Siggelkow [3]). Generally however, simple simulation models are programmed that can only be used for and applied to the particular research topic of the study. For this, an increasing number of researchers turn to preexisting modelling and simulation tools (environments) that support the construction of simulations (e.g. RePast for agent-based systems [8]), which is represented by the bottom right quadrant (round dots, red). Very few simulation research projects cover the additional area marked in the dotted line in the bottom left (long dashes, green). Most researchers do not build an entire simulation environment that can be used for many different research topics and purposes, far beyond the subject of the immediate research needs. This is a pity because future research projects studying different topics could also employ the simulation environment created to derive new

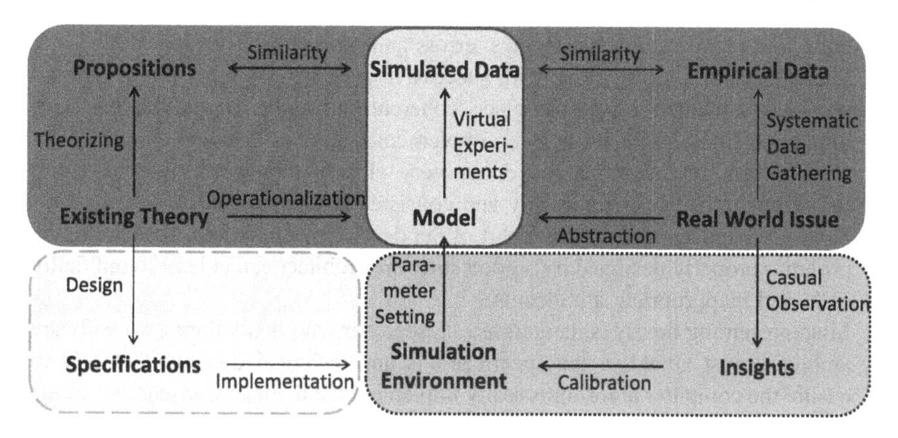

Fig. 6.7 Simulation research activities

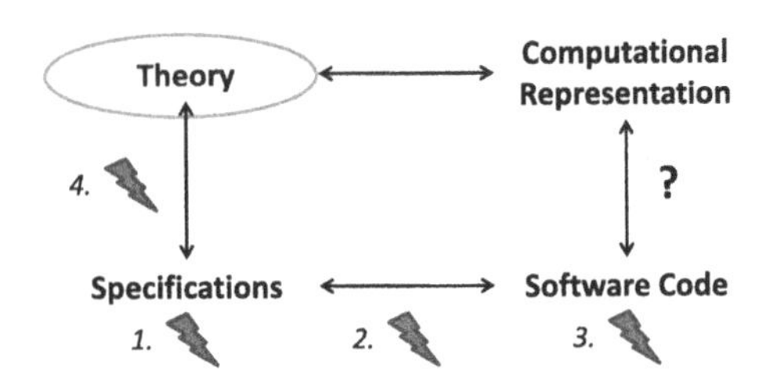

Fig. 6.8 The long road to a simulation environment

theory-driven hypotheses that are empirically testable and which range across numerous applications.

Designing and implementing a full simulation environment or platform (including a simulation execution and reporting suite) that can be used for a variety of research projects is a difficult and laborious process (Fig. 6.8). This probably explains why many people shy away from developing them. Problems can occur at various places, especially if researchers are not able to write the actual software code themselves, but have to rely on software engineers who may not necessarily understand the theory fully. In fact, completely debugging a simulation environment can take several years. However, once developed, it benefits the many researchers who are interested in simulation methods but lack the computer science skills that are necessary to build the software. They can use the graphical user interface to easily set up, run and analyze simulations that model their particular research questions.

To help build a simulation environment, some of the lessons learned from difficulties the author encountered during this effort are listed below. Attending to these four points will help simulation researchers get closer to the proper computational representation of the theory they want to model and avoid losing too much time on software development.

1. If the software development project is inherited, or there are multiple authors, the researcher will have to revisit the specifications thoroughly and check whether they are all correct and appropriate, a process which involves uncovering and repairing inconsistencies, errors and omission in the spec.
2. If the researcher has to work with successive generations of programmers, and proper documentation is not in place, coherent knowledge about what has been implemented and how it has been implemented may be lacking. The software development process must be either very closely monitored over its entire development cycle, or rigorously and consistently documented.
3. Software bugs are an inevitable part of the development process, but having the program properly designed by a good software architect can at least avoid faulty code and inappropriate architecture.
4. Misrepresenting theory is dangerous: a researcher who is not their own software architect must attend to the specifications and technical document closely to ensure the computer algorithms really implement the theory the researcher wants to model.

6.8 Conclusion

This chapter describes our unique research architecture and discusses the applicability of its framework, and looks ahead to many future research projects that could be structured using this approach. Being able to navigate the research space presented by this framework is a first step in using simulation methods to produce novel theory. We hope that the framework will help other researchers conceptualize the different tasks required to realize good simulation studies.

References

1. Carley KM (2002) Computational organizational science and organizational engineering. Simul Model Pract Theory 10(5–7):253–269
2. Christensen CM, Carlile P, Sundahl D (2003) The process of theory-building. Retrieved 01.06.2004, 2004, from http://www.innosight.com/template.php?page=research#Theory%20Building.pdf
3. Csaszar FA, Siggelkow N (2010) How much to copy? determinants of effective imitation breadth. Organ Sci 21(3):661–676
4. Davis JP, Eisenhardt KM, Bingham CB (2007) Developing theory through simulation methods. Acad Manag Rev 32(2):480–499
5. Gilbert N (2008) Agent-based models. Sage Publications, London
6. Gilbert N, Troitzsch KG (2005) Simulation for the social scientist. Open University Press, Maidenhead
7. Harrison JR, Lin Z, Carroll GR, Carley KM (2007) Simulation modeling in organizational and management research. Acad Manag Rev 32(4):1229–1245
8. North MJ, Collier NT, Vos JR (2006) Experiences creating three implementations of the repast agent modeling toolkit. ACM Trans Model Comput Simul 16(1):1–25

Chapter 7
A Ship Motion Short-Term Time Domain Simulator and Its Application to Costa Concordia Emergency Manoeuvres Just Before the January 2012 Accident

Paolo Neri, Mario Piccinelli, Paolo Gubian and Bruno Neri

Abstract In this chapter, we will present a simple but reliable methodology for short-term prediction of a cruise ship behaviour during manoeuvres. The methodology is quite general and could be applied to any kind of ship, because it does not require the prior knowledge of any structural or mechanical parameter of the ship. It is based only on the results of manoeuvrability data contained in the Manoeuvring Booklet, which in turn is filled out after sea trials of the ship performed before his delivery to the owner. We developed this method to support the investigations around the Costa Concordia shipwreck, which happened near the shores of Italy in January 2012. It was then validated against the data recorded in the "black box" of the ship, from which we have been able to extract an entire week of voyage data before the shipwreck. The aim was investigating the possibility of avoiding the impact by performing an evasive manoeuvre (as ordered by the Captain some seconds before the impact, but allegedly misunderstood by the helmsman). The preliminary validation step showed a good matching between simulated and real values (course and heading of the ship) for a time interval of a few minutes. The fact that the method requires only the results registered in the VDR (Voyage Data Recorder) during sea trial tests, makes it very useful for several applications. Among them, we can cite forensic investigation, the development of components

P. Neri (✉)
University of Pisa, Largo L. Lazzarino n.2, 56122 Pisa, Italy
e-mail: paolo.neri@dici.unipi.it

M. Piccinelli · P. Gubian
University of Brescia, Via Branze 38, 25123 Brescia, Italy
e-mail: mario.piccinelli@gmail.com

P. Gubian
e-mail: paolo.gubian@unibs.it

B. Neri
University of Pisa, Via Caruso 16, 56122 Pisa, Italy
e-mail: b.neri@iet.unipi.it

K. Al-Begain and A. Bargiela (eds.), *Seminal Contributions to Modelling and Simulation*, Simulation Foundations, Methods and Applications,
DOI 10.1007/978-3-319-33786-9_7

for autopilots, the prediction of the effects of a given manoeuvre in shallow water, the "a posteriori" verification of the correctness of a given manoeuvre and the use in training simulators for ship pilots and masters.

Keywords Simulation model · Ship manoeuvre · Costa concordia

7.1 Introduction

The problem of simulating the ship motion under the effect of propellers and external forces (wind and drift are the most important) is a central question in the field of naval engineering and training centres for ship pilots and bridge officers. In [1] two different typologies of ship manoeuvring simulation models are described, together with the guidelines for their validation. The distinction is made between models for prediction of ship manoeuvrability (predictive models) and models for ship simulators (simulator models). The targets are quite different.

In the first case, "...prediction of standard ship manoeuvres is needed at the design stage to ensure that a ship has acceptable manoeuvring behaviour, as defined by the ship owner, IMO (International Maritime Organization) or local authorities".

In the second one "Simulator, or time-domain, models are used in real-time, man-in-the-loop simulators, or fast-time simulators for training of deck officers or investigation of specific ships operating in specific harbours or channels".

The first type of model is generated before the ship is built with the aim to foresee, at design stage, some typical manoeuvring characteristics defined by IMO, such as the radius of turning circles or the parameters of Zig-Zag manoeuvres.

The second type of model can be realized after the ship is built up and some mandatory manoeuvrability tests have been performed. This type of model is capable to operate in time domain and can be utilized in simulators for the training of pilots or for other purposes, like that described in this chapter.

For simulator models, a sort of "black-box" approach can be used in which the set of parameters needed in the equations of motion is directly extracted from the experimental data.

Models of the class of so-called black box are nowadays more and more widely employed in many different areas of application. Development of a black box model often starts with a listing of external and internal factors which might have an influence on the behaviour of the system under consideration; then, a mathematical style of modelling is chosen (see below) and a model is written consisting of a number of equations. This model can be more or less domain dependent, in the sense that its equations sometimes might not closely resemble the ones that are usually employed to describe the system in the classical case, and instead take the form of a general mathematical expression such as a quadratic form. After this phase, the coefficients of the equations are determined by means of some fitting process, which makes use of experimental data collected on the real system.

The procedures that are employed to determine the coefficients are almost always well established: in their simplest form they make use of interpolation theory. Different choices of basis functions to represent the model (such as, for example, polynomials, exponentials of Gaussian functions, general neural expressions) present varying degrees of flexibility with respect to the ability of the final model to describe some or all the system properties and behaviours [2]. Recently, approximation theories such as neural network approximation or machine learning approaches in general have been shown to exhibit robust properties in the modelling of complex phenomena and systems [3]. Unfortunately, all of the approaches that lead to a black box model show a tendency to diverge from the real behaviour of the system when users try to use them outside the interval in which the coefficients have been fitted. When employed to predict the behaviour of the system in a domain for which no data have been provided in the fitting phase, the simulated system exhibit a performance that quickly degrades. There is not much that can be done to overcome this degradation, or divergence between simulated and real results, but the onset of such divergence can in many cases be significantly delayed by making use, in the ubiquitous case of a model consisting of ordinary or partial differential equations, of the so-called stiffly-stable methods, which are well known in the literature and which try to make a much better estimate of the discretization error at the expense of a significantly greater computational effort [4, 5].

In the present application, the experimental data used are the results of the manoeuvrability tests performed before the boatyard delivery of the ship to the owner. Several international organizations are qualified for granting the manoeuvrability certificate (the so-called Manoeuvring Booklet); among them, the CETENA S.p.a. is the Italian Company of Fincantieri Group in charge of certifying the compliance of ship sea trials with IMO regulation.

There are a lot of papers, reports, researches available in the literature concerning predictive models which are very important for ship designers and naval architects, as well as for boatyards [6–12]. Conversely, there is very little material concerning time domain simulators of ship motion [13–16]. Here, we will briefly comment two of them.

In [15], after extracting the equations of motion in which the forces and moments acting on hull appear together with those induced by propellers and rudders, the authors perform some numerical simulations. Anyway, the authors do not present any form of validation of the simulator, by means of comparison between simulated and experimental data. In [13], the authors describe a true real-time simulator, a software tool that simulates the motion of the ship and could be used, for instance, for pilots and masters training. The approach value amounts to about 375 m. These results can be compared with those reported in Fig. 7.2 to give a used by the authors is very similar to that of this work. In fact, as the authors say, "The simulation system consists of a ship motion prediction system using a few model parameters which can be evaluated by means of standard ship manoeuvring test or determined easily from databases such as Lloyd's register." In the paper, the validation is performed by the sea trials of the oil tanker ESSO Osaka (turning circle), but the simulations result in relevant errors: the simulated turning circle radius is about

550 m while the measured preliminary idea of the reliability of the simulator presented in this chapter.

We can conclude that, as far as the authors know, there is no evidence in literature of a real-time simulator with reliable behaviour and an extensive characterization of the errors.

After this introduction, in Sect. 7.2 of this chapter, we will describe the data used to extract the model parameters and to validate the reliability of the simulator. Then, in Sect. 7.3, we will describe the motion equations which allowed us to calculate the effect of the forces (internal and external) applied to the hull, whereas in Sect. 7.4 we will present some preliminary results. Finally, Sect. 7.5 contains some conclusions together with a description of future improvements of the model and a list of possible applications.

7.2 Manouvering Booklet and Voyage Data Recorder (VDR) Data

In this section, a brief description of the experimental data used in this work is given. These data were used to extract the model parameters and to validate the simulator by estimating the entity of the errors between simulated and experimental data.

7.2.1 Manoeuvring Booklet

The manoeuvring booklet of a ship reports the results of sea trials performed by the ship before its delivery. Some mandatory tests have to be performed, such as the ones listed in the IMO resolution A—160 (es IV) of 27/11/1968 and following release. The main tests are:

(1) "Turning test", used to determine the effectiveness of rudder to produce steady-state turning (Fig. 7.2).
(2) "Free stop", carried out by suddenly halting a fast-moving ship and measuring the space needed to come to a full stop.
(3) "Crash-stop", similar to the previous one but also including some measures to come to a stop in a shorter time (for example by backing power).
(4) "Zig-Zag manoeuvre", performed by moving the rudder from central position to an assigned angle (usually 10° or 20°) alternatively to starboard and to port (Fig. 7.3).
(5) "Williamson turn", the manoeuvre performed to recover a man overboard (Fig. 7.4).

The sea trials were confined to the area covered by the Differential GPS test range in order to obtain accurate tracking information. This way the position error is reduced to a maximum of 2 m.

During the trials, the position, heading (measured with a gyrocompass), rudder angles, propeller revolution per minute (Costa Concordia had two rudders and two propellers) and shaft power were recorded with a sampling frequency of 1 Hz and stored for further elaborations.

7.2.2 VDR Data

Voyage Data Recorders (VDRs) are systems installed on modern vessels to preserve details about the ship's status, and thus provide information to investigators in case of accident. The ongoing data collection is performed by various devices, and the actual recording of this information is entrusted to an industrial grade computer [17].

The use of VDRs on ships is subjected to the regulations contained in Chap. 5 on "Safety of Navigation" of the "International Convention for the Safety of Life at Sea" (SOLAS). This chapter has been amended in 1999 to adopt the IMO resolution A.861(20) "Performance Standards for Shipborne Voyage Data Recorders (VDRs)" [18]. These regulations, entered into force on July 1, 2002, specify the kinds of ships that are required to carry Voyage Data Recorders; all in all, the list encompasses almost any medium-to-bigger sized ships currently at sea.

The IMO resolution also sets requirements about the operation of the VDR and the kind of information it is required to store. Among the mandatory list, most interesting elements for our analysis are surely the position, speed and heading of the ship, the date/time referenced to UTC, the radar data and the ship automation data, such as the speed of the propellers and the position of the rudder(s).

These informations are collected from a large network of different sensing devices scattered all around the ship. The core of the VDR system is the "concentrator", usually an industrial grade computer, which collects all the data and stores it in a digital memory. The concentrator is powered by the ship's electrical system and also sports a dedicated power source which allows it to work at least 2 h after the loss of main power. At last, a copy of the last 12 h of recording is also maintained inside the digital memory of the FRM (Final Recording Medium), a rugged capsule designed to survive an accident and thus be recovered by the investigators.

7.3 Mathematical Model

The aim of the model is to correlate the input given by the pilot (i.e. motors' power and rudders angles) to the output of the system (i.e. ship's position and orientation). The really complex physical system of the ship has been simplified and schematized with a three-Degrees-Of-Freedom (DOF) model which is adequate for our purpose and consists of the short-term prediction of the trajectory of the ship [19]. Another simplifying assumption is that the effect of wind and current are negligible for short-term simulations. However, these effects have been studied and modelled in the past and could be taken into account in a future improved version of the simulator. Figure 7.1 represents the chosen DOFs (est, nord coordinates and heading ϑ) and their meaning. Using the model of Fig. 7.1, z translation and pitch and roll angles are neglected: those DOFs are crucial for ship's stability studies, but are not relevant for positioning estimation.

Starting from this simple model, we could write the three Newton equations that correlate the accelerations along the three DOFs with the forces acting on the ship:

$$\ddot{\vartheta} = \frac{M_z}{I_\vartheta} = \frac{1}{I_\vartheta}(M_\delta + M_r + M_P) \tag{7.1}$$

$$\ddot{x} = \frac{F_x}{m} = \frac{1}{m}(F_u + F_P + F_{\vartheta,u}) \tag{7.2}$$

$$\ddot{y} = \frac{F_y}{m} = \frac{1}{m}(F_v + F_{\vartheta,v}) \tag{7.3}$$

where M_δ, M_r, M_P represent the torque made by the rudders, the water viscosity and the propellers, respectively. F_u, F_v represent the viscous force made by the water in x and y directions, respectively. F_P represents the force made by the propellers. $F_{\vartheta,u}$, $F_{\vartheta,v}$ are the apparent forces (e.g. centrifugal force) due to the ship rotational speed ($\dot{\vartheta} = r$). I_ϑ represents the moment of inertia along the z axis and m represents the ship's mass.

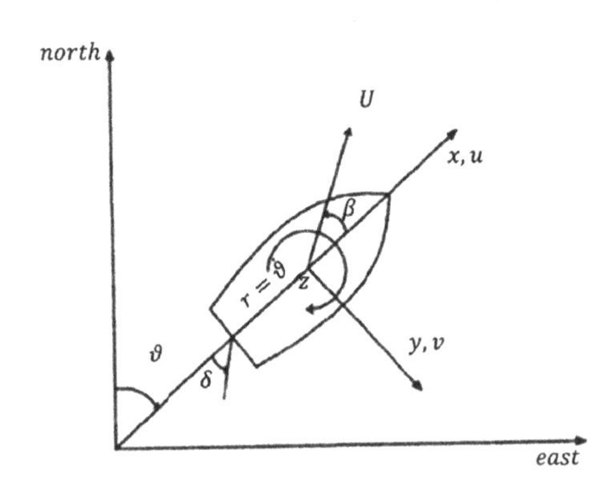

Fig. 7.1 Reference frame

The exact knowledge of those quantities, and an accurate calculation "a priori", would need a deep study of the ship geometry and of the water conditions [20]. Anyway, this work is not aimed to predict the ship behaviour starting from design parameters. Since a sufficient amount of manoeuvring data is available from the Manoeuvring Booklet (CETENA files in the case of Costa Concordia) and from the ship's VDR, it is possible to estimate the relation between the main quantities with an interpolation algorithm.

Equations (7.1), (7.2) and (7.3) can then be rewritten in these simple expressions:

$$\ddot{\vartheta} = N_1 \sin(\delta)\, u\,|\,u\,| + N_2 r\,|\,r\,| + N_3(p_1\,|\,p_1\,| - p_2\,|\,p_2\,|) \tag{7.4}$$

$$\dot{u} = X_1(\mathrm{p}_1\,|\,\mathrm{p}_1\,| + \mathrm{p}_2\,|\,\mathrm{p}_2\,|) + X_2 u\,|\,u\,| + X_3 r^2 + X_4 \ddot{\vartheta} \tag{7.5}$$

$$\dot{v} = Y_1 \mathrm{v}\,|\,\mathrm{v}\,| + Y_2 \ddot{\vartheta} + Y_3 r^2 \tag{7.6}$$

In Eqs. (7.4), (7.5) and (7.6), $\mathrm{N_i}$, $\mathrm{X_i}$, $\mathrm{Y_i}$ represent constant parameters which are estimated starting from the CETENA data.

These equations give a simplified expression of the more complex relations between the relevant quantities. Indeed, they could not be used in an "a priori" model, but the simplification introduced by the use of these expressions is compensated by the constant parameters extracted with an interpolation algorithm. Thanks to these parameters, the model is able to learn from known data how the ship would behave answering to certain input.

7.3.1 Parameters Extraction

The constant parameters needed to integrate the equation of motion (Eqs. (7.4), (7.5) and (7.6)) were obtained from sea trials results collected in CETENA files. To do that, it is useful to consider as an example Eq. (7.4), which can be rewritten to highlight the time-varying terms.

Given that in the CETENA data all the time-varying quantities are known, while the constant parameters are unknown, it is possible to rewrite Eq. (7.4) in a matrix form:

$$\begin{bmatrix} \ddot{\theta}(t_1) \\ \cdots \\ \ddot{\theta}(t) \\ \cdots \\ \ddot{\theta}(t_e) \end{bmatrix} = \begin{bmatrix} \sin(\delta(t_1))u(t_1)^2 r(t_1)^2 \left(p_1(t_1)^2 - p_2(t_1)^2\right) \\ \cdots \\ \sin(\delta(t))u(t)^2 r(t)^2 \left(p_1(t)^2 - p_2(t)^2\right) \\ \cdots \\ \sin(\delta(t_e))u(t_e)^2 r(t_e)^2 \left(p_1(t_e)^2 - p_2(t_e)^2\right) \end{bmatrix} * \begin{bmatrix} N_1 \\ N_2 \\ N_3 \end{bmatrix} \tag{7.7}$$

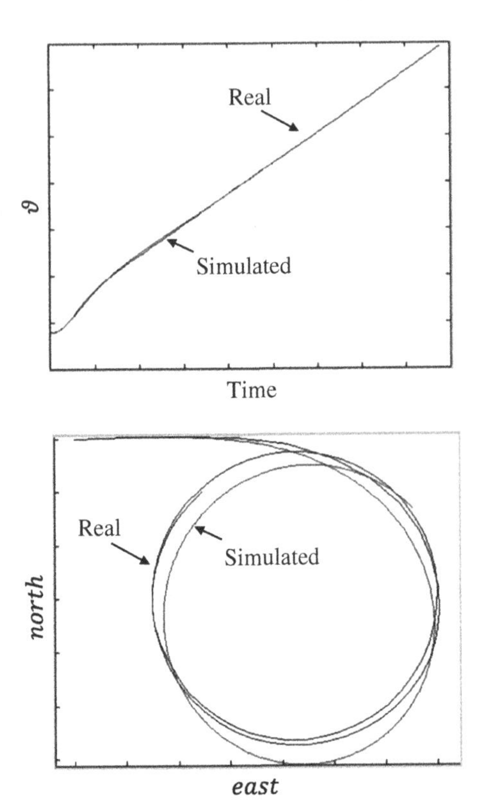

Fig. 7.2 Turning circle: simulation time 500 s

In this expression, the number of equations is much greater than the number of unknown terms, so that the system can be solved in terms of a least square interpolation perform a pseudo-inverse of the matrix (*pinv* function). The same procedure can be repeated for the other two equations of motion (Eqs. (7.5) and (7.6)) in order to calculate all the constant parameters needed (i.e. N_i, X_i, Y_i). The manoeuver described in CETENA records used for this calculation was the turning circle (Fig. 7.2).

7.3.2 *Initial Conditions*

Once the parameters have been estimated, it was possible to integrate Eqs. (7.4), (7.5) and (7.6) to obtain a simulation of the ship behaviour. The integration method used was the Forward Euler method. The simulator needs as an input the pilot's instruction (i.e. rudder angles and propellers speed) and the initial conditions (i.e. starting coordinates and heading and starting linear and rotational speed). The speed information could be obtained by deriving coordinates and heading with respect to the time. Anyway, those data are affected by the error caused by the GPS and the compass sensitivity, so that a correct estimation of the initial speed is not trivial.

A simple incremental ratio would give an unstable result, considering two consequent points, because of the measures noise. A more robust way to proceed is using a polynomial interpolation to approximate the position and the heading data, in order to filter out such spurious effects.

7.3.3 Validation

The best way to validate the aforementioned algorithm is to perform simulations of known data, in order to compare the simulated results with real ones. This was done with CETENA data contained in the Manoeuvring Booklet. The IMO standard manoeuvres were simulated by giving to the simulator the same values of rudder angles and propeller rpm (round per minute) used in real manoeuvres. The real data (position and heading) recorded by the sensors used during the sea trials were then compared with the simulator output. In Figs. 7.2, 7.3 and 7.4 the results of this comparison are shown for turning circle, zig-zag and Williamson turn manoeuvres, respectively. The results are surprisingly good if we take into account the simplicity of the mathematical/physical model used and the few information required to build it (i.e. the set of data collected during the sea trials).

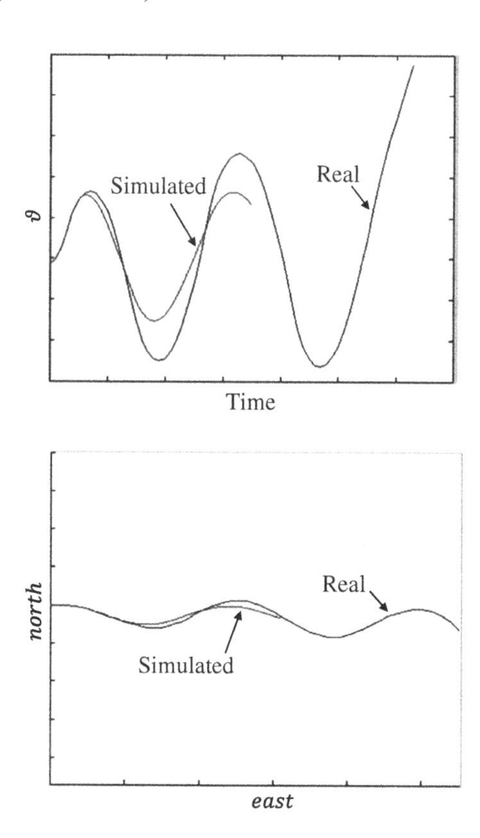

Fig. 7.3 Zig-Zag: simulation time 300 s

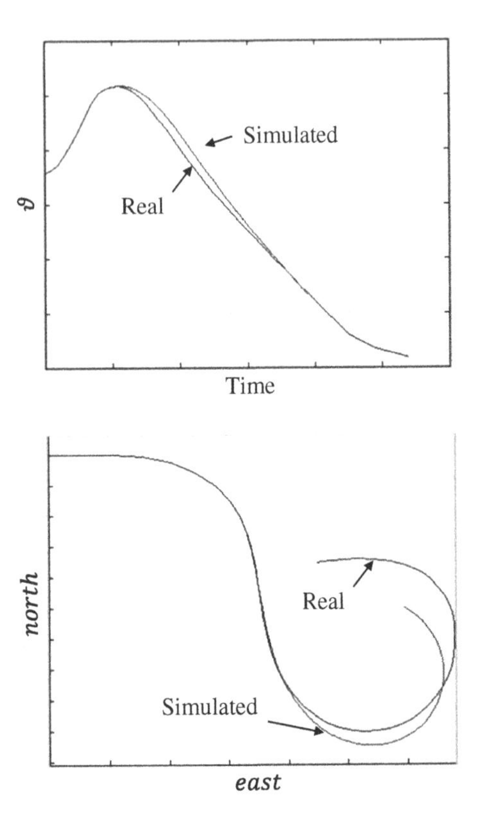

Fig. 7.4 Williamson turn: simulation time 300 s

To better prove the reliability of the simulator, a further validation was performed by using the data recorded in the VDR of Costa Concordia during the week before the shipwreck. It must be noted that the real-time simulator had been realized for short-term simulation, and that the final goal was the simulation of the manoeuvres ordered by the Master of the Costa Concordia just before the impact with the rocks.

Three events of the navigation recorded in the VDR were selected for validation. We will call them Palamos Turn, Palma Zig-Zag, Giglio Zig-Zag, because of the site in which they occurred and the manoeuvre typology. It is worth to underline that Giglio Zig-Zag manoeuvre ends with the impact against the rocks. Comparison results are reported in Table 7.1, in terms of position and heading errors. The errors were calculated at different simulation times.

As can be seen, the simulation results are reliable for a simulation time of about 60 s, while the error increases as the simulation time reaches 120 s. This can be justified by the fact that in a short time simulation, environmental conditions, neglected in the model, play a second order role in ship behaviour, while become much more important as the simulation time increases. Furthermore, the integration method used is really simple in order to reduce simulation time, then the effects of the errors cumulate as simulation time increases.

Table 7.1 Simulation errors

Palamos Turn			
Sim. time (s)	*East* error (m)	*North* error (m)	Heading error (°)
30	1.6	0.3	−0.76
60	15.9	2.9	−3.42
120	82.2	−40.8	−14.60
Palma Zig-Zag			
Sim. time (s)	*East* error (m)	*North* error (m)	Heading error (°)
30	−0.5	−0.8	−0.07
60	−1.5	−2.1	−0.09
120	−4.1	−7.6	−0.21
Giglio Zig-Zag			
Sim. time (s)	*East* error (m)	*North* error (m)	Heading error (°)
30	5.3	5.2	−0.26

7.4 Results

In the case of Giglio Zig-Zag, the simulated position of the ship at 21.45.11 (the impact time and the end of the simulation time) is considered. This simulation begins 30 s before the impact, that is, more or less the instant at which the Master saws the rocks and made the extreme attempt to avoid the impact by ordering a "Zig-Zag" manoeuvre. Apparently, the helmsman did not understand the orders of the master and made a mistake, by putting the rudder hard to starboard instead of hard to port as he was ordered. At this point, we repeated the simulation by giving to the simulator the correct position of the rudders as ordered. The timing of the order and the order itself were found in the recordings of the bridge audio, which was stored in the VDR, according with IMO Regulation. The result is shown in Fig. 7.5, where the grey line represents the real position of the ship at the impact time and the black dashed one indicates the simulated position of the ship without the error of the helmsman. The black ellipse in the figure represents the rock. The quite surprising conclusion is that the impact point at the left side of the ship in the simulation is more or less 10 m far from the rock. Taking into consideration the sea bottom shape at the accident site, this distance would be enough to allow the ship to pass next to the rock without any damage.

To complete the discussion, we also considered the error worst case. The maximum observed value of the error after 30 s of simulation is about 7 m: in this case, we would still have an impact but the actual impact point would have been about 18 m behind the actual one. It is our opinion that, in this worst case, at least one of two engine rooms (more likely both of them) would have been left unscathed by the accident, and the ship would have been capable to stay afloat. In this case, the final consequences on the ship and the passengers would have been really different.

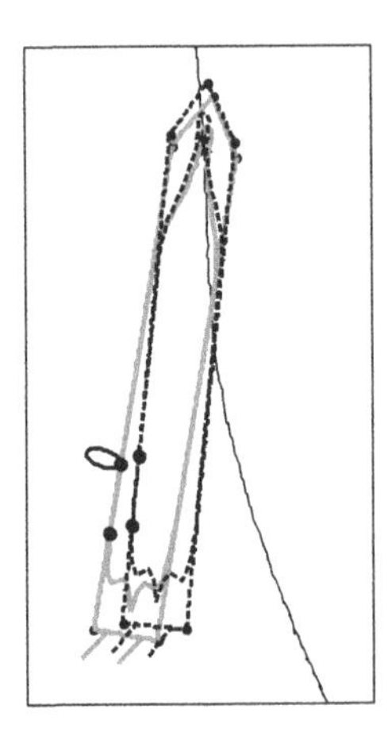

Fig. 7.5 Real (*grey*) and simulated without the helmsman's error (*black-dashed*) position of the ship at the instant of the impact

7.5 Conclusion

A simple time domain short-term simulator of the motion of a ship in free water has been presented. The simulator is based on a quite simple "black-box" model that does not require the knowledge of structural and mechanical characteristics of the ship. The position of the ship and its orientation versus time are calculated by using a mathematical model in which ten free parameters are used. These parameters are bound to the manoeuvrability characteristics of the ship, and can be obtained by using the recorded results of sea trials performed with the ship before its delivery.

Despite the great number of papers in the literature dealing with the problem of simulating the manoeuvres of a specific ship, just a few of them deal with the problem of a time domain algorithm able to simulate, at least for a short time, the effect of propellers and rudders during a manoeuvre.

Our simulator could be used for pilots and masters training, in autopilot systems, in emergency manoeuvre foreseeing and so on. Further improvements will be introduced in the model in order to take into account the effects of wind and stream.

In the second part of the paper, the procedure has been used to simulate the Costa Concordia cruise ship behaviour, in order to study the effect of the allegedly wrong manoeuvre performed by the helmsman just before the impact. The result was that if the helmsman would properly execute the Master's orders, the impact could have been avoided or, at least, by considering the worst case, its effects could have been much less devastating. Finally, to definitely confirm these conclusions, the Concordia's twin ship, Costa Serena, could be used to perform a sea trial in which the manoeuvres with and without the helmsman's error should be executed. The different tracks recorded during the test could then be compared to evaluate the actual consequences of the helmsman's error on the Costa Concordia shipwreck.

Acknowledgements The authors are grateful to CODACONS, an Italian consumers' rights association that is supporting some Costa Concordia survivors in the trial, for financial support and legal assistance.

References

1. ITTC (1999) Recommended procedures-testing and extrapolation methods-manoeuvrability validation of manoeuvring simulation models. http://www.simman2008.dk/PDF/7.5–02-06-03%20(simulation%20models).pdf
2. Maffezzoni P, Gubian P (1995) A new unsupervised learning algorithm for the placement of centers in a radial basis function neural network (RBFNN). In: Proceedings of 1995 international conference on neural networks, Perth, Western Australia, vol 3, pp 1258–1262
3. Poggio T, Girosi F (1990) Networks for approximation and learning. In: Proceedings of the IEEE, vol 78, no 9, pp 1481–1497
4. Gear CW (1971) Numerical initial value problems for ordinary differential equations. Prentice-Hall
5. Daldoss L, Gubian P, Quarantelli M (2001) Multiparameter time-domain sensitivity computation. IEEE Trans Circuits Syst I. Fundam Theory Appl 48:1296–1307
6. Ishiguro T, Tanaka S, Yoshimura Y (1996) A study on the accuracy of the recent prediction technique of ship's manoeuvrability at an early design stage. In: Chislett MS (ed) Proceedings of marine simulation and ship manoeuvrability, Copenhagen, Denmark
7. ITTC (1996) International Towing Tank Conference, 1996, Manoeuvring Committee—Final Report and Recommendations to the 21st ITTC. In: Proceedings of 21st ITTC, Bergen, Trondheim, Norway, vol 1, pp 347–398
8. ITTC (2002) International Towing Tank Conference, 1999, "Manoeuvring Committee—Final Report and Recommendations to the 22nd ITTC". In: Proceedings of 22nd ITTC, Seoul/Shanghai. International Towing Tank Conference
9. ITTC (2002) Esso Osaka Specialist Committee—Final Report and Recommendations to the 23rd ITTC. In: Proceedings of 23rd ITTC, Venice, Italy
10. Oltmann P (1996) On the influence of speed on the manoeuvring behaviour of a container carrier. In: Proceedings of marine simulation and ship manoeuvrability, Copenhagen, Denmark. ISBN 90 54108312
11. Revestido E et al (2011) Parameter estimation of ship linear maneuvering model. In: Proceedings of Ocean 2011, Santander, Spain, pp 1–8, 6–9
12. Sutulo S, Moreira L, Guedes Soares C (2002) Mathematical models for ship path prediction in manoeuvring simulation systems. In: Ocean Engineering, vol 29, pp 1–19. Pergamon Press
13. Sandaruwan D, Kodikara N, Keppitiyagama C, Rosa R (2010) A six degrees of freedom ship simulation system for maritime education. Int J Adv ICT Emerg Regions 34–47
14. Gatis B, Peter FH (2007) Fast-time ship simulator. Safety at Sea
15. Mohd NCW, Wan NWB, Mamat M (2012) Ship manoeuvring assessment by using numerical simulation approach. J Appl Sci Res 8(3):1787–1796
16. Shyh-Kuang U, Lin D, Liu C-H (2008) A ship motion simulation system. Virtual reality. Springer, pp 65–76
17. Piccinelli M, Gubian P (2013) Modern Ships voyage data recorders: a forensics perspective on the costa concordia shipwreck. Digit Investig 10. Elsevier
18. IMO—International Maritime Organization. Resolution A.861(20) (1997) Performance standards for shipborne Voyage Data Recorders (VDRs). Accessed Nov 27
19. Maimun A, Priyanto A, Muhammad AH, Scully CC, Awal ZI (2011) Manoeuvring prediction of pusher barge in deep and shallow water. Ocean Eng 38:1291–1299
20. Maimun A, Priyanto A, Sian AY, Awal ZI, Celement CS, Waqiyuddin M (2013) A mathematical model on manoeuvrability of a LNG tanker in vicinity of bank in restricted water. Saf Sci 53:34–44

Author Biographies

Paolo Neri was born in Palermo, Italy, and he earned his Master Degree in Mechanical Engineering at the University of Pisa in 2012. He is now attending the second year of the PhD course at University of Pisa. His main research interests include physical systems simulation, experimental modal analysis, machine dynamics and especially the development of robot-assisted automatic procedures for mechanical systems dynamic characterization. His e-mail address is: paolo.neri@dici.unipi.it.

Mario Piccinelli was born in Lovere, Italy, and earned his Master's Degree in Electronic Engineering at the University of Brescia, Italy, in 2010. He is now a PhD candidate in Information Engineering with a thesis about Digital Forensics. His main research interests include the extraction and analysis of data from modern embedded devices, such as smartphones or eBook readers. He is also working as a consultant for both prosecutors and lawyers for matters related to the analysis of digital evidence, and was appointed as private consultant in the Costa Concordia shipwreck trial. His e-mail address is: mario.piccinelli@gmail.com and his Web page can be found at http://www.mariopiccinelli.it.

Paolo Gubian received the Dr. Ing. degree "summa cum laude" from Politecnico di Milano, Italy, in 1980. He consulted for ST Microelectronics in the areas of electronic circuit simulation and CAD system architectures. During this period, he worked at the design and implementation of ST-SPICE, the company proprietary circuit simulator. From 1984 to 1986, he was a Visiting Professor at the University of Bari, Italy, teaching a course on circuit simulation. He also was a visiting scientist at the University of California at Berkeley in 1984. In 1987, he joined the Department of Electronics at the University of Brescia, Italy where he is now an Associate Professor in Electrical Engineering. His research interests are in statistical design and optimization and reliability and robustness in electronic system architectures. His e-mail address is: paolo.gubian@unibs.it and his Web-page can be found at http://www.ing.unibs.it/~gubian.

Bruno Neri was born in 1956 and received his "Laurea" degree "cum laude" from University of Pisa in 1980. In 1983, he joined the Department of Information Engineering of the same University, where he is full Professor of Electronic since 2000. In recent years, his research activity has addressed the design of radio-frequency integrated circuits for mobile communications and for biomedical applications. Presently, he is a Technical Consultant in the lawsuit for the shipwreck of Costa Concordia cruise ship. Bruno Neri is co-author of more than 100 papers published in peer-reviewed Journals and Proceedings of International Conferences. His e-mail address is: b.neri@iet.unipi.it.

Chapter 8
Fuzzy Modelling and Fuzzy Collaborative Modelling: A Perspective of Granular Computing

Witold Pedrycz

Abstract The study elaborates on current developments in fuzzy modelling, especially fuzzy rule-based modelling, by positioning them in the general setting of granular computing. This gives rise to granular fuzzy modelling where the models built on a basis of fuzzy models are then conceptually augmented to make them in rapport with experimental data. Two main directions of granular fuzzy modelling dealing with distributed data and collaborative system modelling and transfer knowledge are formulated and the ensuing design strategies are outlined.

Keywords System modelling · Fuzzy models · Granular fuzzy models · Granular computing · Information granules

8.1 Introduction

While studying a plethora of methodologies, architectures and algorithms of models encountered in system modelling, there is an apparent conclusion that there are no ideal models as lucidly stated by G.E.P. Box [5]

> Since all models are wrong the scientist cannot obtain a "correct" one by excessive elaboration. On the contrary following William of Occam he should seek an economical description of natural phenomena. Just as the ability to devise simple but evocative models is the signature of the great scientist so overelaboration and overparameterization is often the mark of mediocrity.

W. Pedrycz (✉)
Department of Electrical and Computer Engineering, University of Alberta,
Edmonton, AB T6R 2V4, Canada
e-mail: wpedrycz@ualberta.ca

W. Pedrycz
Faculty of Engineering, Department of Electrical and Computer Engineering,
King Abdulaziz University, Jeddah 21589, Saudi Arabia

W. Pedrycz
Systems Research Institute, Polish Academy of Sciences, Warsaw, Poland

K. Al-Begain and A. Bargiela (eds.), *Seminal Contributions to Modelling and Simulation*, Simulation Foundations, Methods and Applications,
DOI 10.1007/978-3-319-33786-9_8

An issue on how to establish sound and justified trade-offs between accuracy and interpretability (transparency) of models has been high on system modelling agenda for a long time. With the advent of fuzzy sets and fuzzy models constructed with conceptual entities—building blocks formed by means of fuzzy sets and fuzzy relations, one has entered an era of designing models being supported by the technology of fuzzy sets. Fuzzy sets are information granules positioned at a certain level of abstraction. In a natural way they help to establish modelling pursuits and facilitate a formation of the resulting model at a certain level of generality by ignoring unnecessary details and emphasize the essentials of the system or phenomenon to be modelled. Fuzzy sets and fuzzy modelling has been around for about five decades. They have resulted in a plethora of concepts, algorithms, design, development practices and applications.

The objective of this study is to identify and formulate several directions of fuzzy modelling in which fuzzy models are developed in the setting of Granular Computing and address new categories of timely modelling problems such as collaborative fuzzy modelling and transfer knowledge; both of them benefit from the use of information granules.

To make a study self-contained and well focused (given the plethora of fuzzy models, it is beneficial to identify some landmarks of fuzzy modelling), we concentrate on Takagi-Sugeno fuzzy rule-based models. In their generic format, these models come in the form of the rules associating some regions in the input space with local, usually linear, models. The regions in the input space are described in the form of some information granules.

The study is structured in the following way. In Sect. 8.2, we briefly recall the generic version of the fuzzy rule-based models by stressing the design process. Information granules and granular computing are discussed in Sect. 8.3. The fundamentals of granular computing, namely, a principle of justifiable granularity and an optimal allocation of information granularity being at the heart of building granular fuzzy models are presented in Sects. 8.4 and 8.5, respectively. In Sect. 8.6, we elaborate on collaborative fuzzy modelling realized in the presence of a family of data. A problem of transfer knowledge is discussed in Sect. 8.7; it turns out that information granules arise as a sound vehicle to formulate the problem and establish a design strategy.

8.2 Fuzzy Rule-Based Models

Takagi-Sugeno fuzzy rule-based model M is described in the form of collection of rules (conditional "if-then" statements) [1, 2]

$$\text{if } \boldsymbol{x} \text{ is } A_i \text{ then } y \text{ is } f_i(\boldsymbol{x}, \mathbf{a}_i) \tag{8.1}$$

where A_i are fuzzy sets defined in a multidimensional space of reals $\boldsymbol{R}^n$ and f_i is a function describing locally relationship between the input x restricted to A_i and the output, $i = 1, 2, \ldots, c$, dim $(\boldsymbol{x}) = n$. The overall input–output characteristics of the model described by (8.1) are expressed as follows:

$$y = \sum_{i=1}^{c} A_i(\boldsymbol{x}) f_i(\boldsymbol{x}) \tag{8.2}$$

There have been a number of studies devoted to the analysis of fuzzy rule-based models highlighting the main advantages of such models including their modularity and transparency [8, 9]. Fuzzy rule-based models are universal approximators and this speaks loudly to the advantages of these architectures [7, 19, 20].

The design of this rule-based model is realized in the presence of data $\boldsymbol{D} = \{(\boldsymbol{x}_k, target_k)\}$, $k = 1, 2, \ldots, N$. Irrespectively of various development schemes, a majority of them consist of the two design steps, namely,

(i) formation of condition part—fuzzy sets $A_1, A_2, \ldots, A_c$. This is done in unsupervised mode by involving clustering methods. FCM is commonly used here [4]. The essence of this phase is to build information granules in the input space, namely, to determine a structure in the input space with the clusters that are very likely homogeneous information granules over which some simple functions f_i can be regarded as sufficient to model even complex nonlinear relationships between input and output spaces.

(ii) optimization of parameters of the local functions f_i (which are linear). Assuming f_i being linear functions with respect to their parameters the optimal solution (producing a global minimum) when minimizing a sum of squared errors is obtained analytically.

8.3 Information Granules and Granular Computing

To make the study presented here as self-contained and also to establish a better focus of the paper, we present a concise introduction to granular computing [3, 10, 16] regarded as a formal vehicle to cast data analysis and modelling tasks in a certain conceptual framework.

Information granules are intuitively appealing constructs, which play a pivotal role in human cognitive and decision-making activities. We perceive complex phenomena by organizing existing knowledge along with available experimental evidence and structuring them in a form of some meaningful, semantically sound entities, which are central to all ensuing processes of describing the world, reasoning about the environment and support decision-making activities. The term information granularity itself has emerged in different contexts and numerous areas of application. It carries various meanings. One can refer to artificial intelligence in which case information granularity is central to a way of problem solving through problem decomposition where various subtasks could be formed and solved individually. In general, by information granule, one regards a collection of elements drawn together by their closeness (resemblance, proximity, functionality, etc.) articulated in terms of some useful spatial, temporal, or functional relationships. Subsequently, granular computing is about representing, constructing and processing information granules.

We can refer here to some areas, which offer compelling evidence as to the nature of underlying processing and interpretation in which information granules play a pivotal role: image processing, processing and interpretation of time series, granulation of time, design of software systems. Information granules are examples of abstractions. As such, they naturally give rise to hierarchical structures: the same problem or system can be perceived at different levels of specificity (detail) depending on the complexity of the problem, available computing resources and particular needs to be addressed. A hierarchy of information granules is inherently visible in processing of information granules. The level of detail (which is represented in terms of the size of information granules) becomes an essential facet facilitating a way a hierarchical processing of information with different levels of hierarchy indexed by the size of information granules.

Even such commonly encountered and simple examples presented above are convincing enough to lead us to ascertain that (a) information granules are the key components of knowledge representation and processing, (b) the level of granularity of information granules (their size, to be more descriptive) becomes crucial to the problem description and an overall strategy of problem solving, (c) hierarchy of information granules supports an important aspect of perception of phenomena and deliver a tangible way of dealing with complexity by focusing on the most essential facets of the problem, (d) there is no universal level of granularity of information; essentially the size of granules becomes problem-oriented and user dependent.

There are several well-known formal settings in which information granules can be expressed and processed.

Sets (intervals) realize a concept of abstraction by introducing a notion of dichotomy: we admit element to belong to given information granule or to be excluded from it. Along with set theory comes a well-developed discipline of interval analysis. Alternatively, to an enumeration of elements belonging to a given set, sets are described by characteristic functions taking on values in $\{0, 1\}$.

Fuzzy sets [17, 18] provide an important conceptual and algorithmic generalization of sets. By admitting partial membership of an element to a given information granule we bring an important feature which makes the concept to be in rapport with reality. It helps in working with the notions where the principle of dichotomy is neither justified nor advantageous. The description of fuzzy sets is realized in terms of membership functions taking on values in the unit interval. Formally, a fuzzy set A is described by a membership function mapping the elements of a universe $\mathbf{X}$ to the unit interval $[0, 1]$.

Shadowed sets [10] offer an interesting description of information granules by distinguishing among elements, which fully belong to the concept, are excluded from it and whose belongingness is completely *unknown*. Formally, these information granules are described as a mapping X: $\boldsymbol{X} \rightarrow \{1, 0, [0, 1]\}$ where the elements with the membership quantified as the entire [0, 1] interval are used to describe a shadow of the construct. Given the nature of the mapping here, shadowed sets can be sought as a granular description of fuzzy sets where the shadow is used to localize unknown membership values, in which fuzzy sets are distributed over the entire universe of discourse. Note that the shadow produces nonnumeric descriptors of membership grades.

Probability oriented information granules are expressed in the form of some probability density functions or probability functions. They capture a collection of elements resulting from some experiment. In virtue of the concept of probability, the granularity of information becomes a manifestation of occurrence of some elements. For instance, each element of a set comes with a probability density function truncated to [0, 1], which quantifies a degree of membership to the information granule.

Rough sets emphasize a roughness of description of a given concept X when being realized in terms of the indiscernibility relation provided in advance. The roughness of the description of X is manifested in terms of its lower and upper approximations of a certain rough set.

From the perspective of this study, there are two important and directly applicable characterizations of information granules, namely, coverage and specificity [10, 11, 13].

Coverage The concept of coverage of information granule, *cov*(.) is discussed with regard to some experimental data existing in $\boldsymbol{R}^n$, that is $\{\boldsymbol{x}_1, \boldsymbol{x}_2, \ldots, \boldsymbol{x}_N\}$ and as the name stipulates, is concerned with its ability to represent (cover) these data. In general, the larger number of data is being "covered", the higher the coverage measure. Formally, the coverage is a nondecreasing function of the number of data that are represented by the given information granule A. Depending upon the nature of information granule, the definition of $cov(A)$ can be properly expressed. For instance, when dealing with a multidimensional interval (hypercube) A, $cov(A)$ in its normalized form is related with the cardinality of the data belonging to A, $cov(A) = \frac{1}{N}\text{card}(\boldsymbol{x}_k | \boldsymbol{x}_k \in A\}$ For fuzzy sets the coverage is realized as a σ-count of A, where we summed up the degrees of membership of $\boldsymbol{x}_k$ to A, $cov(A) = \frac{1}{N}\sum_{k=1}^{N} A(\boldsymbol{x}_k)$.

Specificity Intuitively, the specificity relates to a level of abstraction conveyed by the information granules. The higher the specificity, the lower the level of abstraction. The monotonicity property holds: if for the two information granules A and B one has $A \subset B$ (when the inclusion relationship is articulated according to the formalism in which A and B are expressed) then specificity, *sp*(.) [15] satisfies the following inequality: $sp(A) \geq sp(B)$. Furthermore for a degenerated information granule comprising a single element $\boldsymbol{x}_0$ we have a boundary condition $sp(\{\boldsymbol{x}_0\}) = 1$. In case of a one-dimensional interval information granules, one can contemplate expressing specificity on a basis of the length of the interval, say $sp(A) = \exp(-\text{length}(A))$; obviously the boundary condition specified above holds here. If the range of the data is available (or it could be easily determined), then the specificity measure can be realized in the following way $sp(A) = 1 - |b - a|/range$ where $A = [a, b]$, $range = [\min_k x_k, \max_k x_k]$.

The realizations of the definitions can be augmented by some parameters that contribute to their flexibility. It is also intuitively apparent that these two characteristics are associated: the increase in one of then implies a decrease in another: an information granule that "covers" a lot of data cannot be overly specific and vice versa.

8.3.1 Information Granules of Higher Type

By information granules of *higher type* (second type and *n*th type, in general) we mean granules in the description of whose we use information granules rather than numeric entities. For instance, in case of type-2 fuzzy sets we are concerned with information granules—fuzzy sets whose membership functions are granular. As a result, we can talk about interval-valued fuzzy sets, fuzzy sets (or fuzzy sets, for brief), probabilistic sets and alike. The grade of belongingness are then intervals in [0, 1], fuzzy sets with support in [0, 1], probability functions truncated to [0, 1], etc. In case of type-2 intervals we have intervals whose bounds are not numbers but information granules and as such can be expressed in the form of intervals themselves, fuzzy sets, rough sets, or probability density functions. Information granules *of higher order* are those whose description is realized over a universe of discourse whose elements are information granules. In some sense rough sets could be sought as information granules of order-2. Information granules have been encountered in numerous studies reported in the literature; in particular stemming from the area of fuzzy clustering [12] in which fuzzy clusters of type-2 have been investigated [6] or they are used to better characterize a structure in the data and could be based upon the existing clusters [5].

8.4 The Principle of Justifiable Granularity

The principle of justifiable granularity [10, 11] offers a conceptual and algorithmic setting to develop an information granule. The principle is general as it shows a way of forming information granule without being restricted to certain formalism in which information granularity is expressed and a way experimental evidence using which this information granule comes from. Let us denote one-dimensional numeric data of interest (for which an information granule is to be formed) by $\mathbf{Z} = \{z_1, z_2, \ldots, z_N\}$. Denote the largest and the smallest element in $\mathbf{Z}$ by z_{min} and z_{max}, respectively. On a basis of $\mathbf{Z}$ we are form an information granule A so that it attempts to satisfy two intuitively requirements of coverage and specificity. The first one implies that the information granule is justifiable, namely, it embraces (covers) as many elements of $\mathbf{Z}$ as possible. The second one is to assure that the constructed information granule exhibits a well-defined semantics by being specific enough. When constructing a fuzzy set, say a triangular one, we first start with a numeric

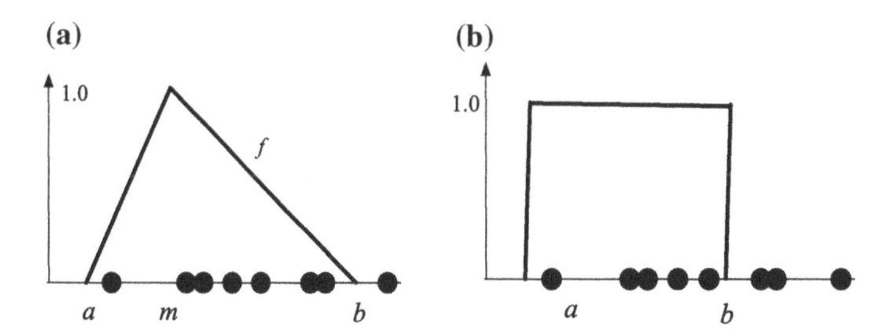

Fig. 8.1 Design of information granules with the use of the principle of justifiable granularity **a** triangular membership function, **b** interval (characteristic function)

representative of **Z**, say a mean or a modal value (denoted here by *m*) and then separately determine the lower bound (*a*) and the upper bound (*b*). In case of an interval *A*, we start with a modal value and then determine the lower and upper bound. Refer to Fig. 8.1.

The construction of the bounds is realized in the same manner for the lower and upper bound so in the following we describe only a way of optimizing the upper bound (*b*).

The coverage criterion is expressed as follows:

$$cov(A) = \sum_{x_k : x_k \in [m,b]} g(z_k) \tag{8.3}$$

where *g* is a decreasing linear portion of the membership function. For an interval form of A the coverage is a count of the number of data included in [m, b],

$$cov(A) = \text{card}\{z_k | z_k \in [m, b]\} \tag{8.4}$$

The above coverage requirement states that we reward the inclusion of z_i in *A*. The specificity *sp*(*A*) is one of those specified in the previous section. As we intend to maximize coverage and specificity and these two criteria are in conflict, an optimal value of b is the one, which maximizes the product of the two requirements

$$Q(b) = cov(A) * sp(A)^{\xi} \tag{8.5}$$

Furthermore, the optimization performance index is augmented by an additional parameter ξ assuming nonnegative values. It helps control an impact of the specificity in the formation of the information granule. The higher the value of ξ, the more essential is its impact on *A*. If $\xi = 0$ the only criterion of interest is the coverage. Higher values of ξ underline the importance of specificity; as a result *A* is more specific. The result of optimization comes in the form $b_{\text{opt}} = \arg\max_b Q(b)$.

8.5 An Emergence of Granular Models: Structural Developments

The concept of the granular models forms a generalization of numeric models no matter what their architecture and a way of their construction are. In this sense, the conceptualization offered here are of general nature. They also hold for any formalism of information granules. A numeric model M_0 constructed on a basis of a collection of training data ($\boldsymbol{x}_k$, target$_k$), $\boldsymbol{x}_k \in \boldsymbol{R}^n$ and $target_k \in \boldsymbol{R}$ comes with a collection of its parameters $\boldsymbol{a}_{\text{opt}}$ where $\boldsymbol{a} \in \boldsymbol{R}^p$. Quite commonly, the estimation of the parameters is realized by minimizing a certain performance index Q (say, a sum of squared error between $target_k$ and $M_0(\boldsymbol{x}_k)$), namely, $\boldsymbol{a}_{\text{opt}}$ = arg Min$_{\boldsymbol{a}}$ Q($\boldsymbol{a}$). To compensate for inevitable errors of the model (as the values of the index Q are never equal identically to zero), we make the parameters of the model information granules, resulting in a vector of information granules $\boldsymbol{A} = [A_1\ A_2\ \ldots\ A_p]$ built around original numeric values of the parameters $\boldsymbol{a}$. The elements of the vector $\boldsymbol{a}$ are generalized, the model becomes granular and subsequently the results produced by them are information granules. Formally speaking, we have

- granulation of parameters of the model $\boldsymbol{A} = G(\boldsymbol{a})$ where G stands for the mechanisms of forming information granules, namely, building an information granule around the numeric parameter
- result of the granular model for any $\boldsymbol{x}$ producing the corresponding information granule Y, $Y = M_1(\boldsymbol{x}, \boldsymbol{A}) = G(M_0(\boldsymbol{x})) = M_0(\boldsymbol{x}, G(\boldsymbol{a}))$.

Information granulation is regarded as an essential design asset. By making the results of the model granular (and more abstract in this manner), we realize a better alignment of $G(M_0)$ with the data. Intuitively, we envision that the output of the granular model "covers" the corresponding target. Formally, let cov($target$, Y) denote a certain coverage predicate (either Boolean or multivalued) quantifying an extent to which target is included (covered) in Y.

The design asset is supplied in the form of a certain allowable level of information granularity ε which is a certain nonnegative parameter being provided in advance. We allocate (distribute) the design asset across the parameters of the model so that the coverage measure is maximized while the overall level of information granularity serves as a constraint to be satisfied when allocating information granularity across the model, namely, $\sum_{i=1}^{p} \varepsilon_i = p\varepsilon$. The constraint-based optimization problem reads as follows:

$$\max_{\varepsilon_1,\varepsilon_2,\ldots,\varepsilon_p} \sum_{k=1}^{N} \text{cov}(target_k \in Y_k)$$

subject to

$$\sum_{i=1}^{p} \varepsilon_i = p\varepsilon \text{ and } \varepsilon_i \geq 0 \tag{8.6}$$

The monotonicity property of the coverage measure is apparent: the higher the values of ε, the higher the resulting coverage. Hence the coverage is a nondecreasing function of ε.

Along with the coverage criterion, one can also consider the specificity of the produced information granules. It is a nonincreasing function of ε. The more general form of the optimization problem can be established by engaging the two criteria leading to the two-objective optimization problem-determine optimal allocation of information granularity $[\varepsilon_1, \varepsilon_2, \ldots, \varepsilon_p]$ so that the coverage and specificity criteria become maximized.

Plotting these two characteristics in the coverage-specificity coordinates offers a useful visual display of the nature of the granular model and possible behaviour of the behaviour of the granular model as well as the original model. Several illustrative plots, as shown in Fig. 8.1, illustrate typical changes in the specificity/coverage when changing the values of information granularity ε. One can consider those coming as a result of the maximization of coverage while reporting also the obtained values of the specificity. There are different patterns of the changes between coverage and specificity. The curve may exhibit a monotonic change with regard to the changes in ε and could be approximated by some linear function. There might be some regions of some slow changes of the specificity with the increase of coverage with some points at which there is a substantial drop of the specificity values. A careful inspection of these characteristics helps determine a suitable value of ε—any further increase beyond this limit might not be beneficial as no significant gain of coverage is observed however the drop in the specificity compromises the quality of the granular model. Furthermore, Fig. 8.2 highlights an identification of suitable values of the level of information granularity.

The global behaviour of the granular model can be assessed in a global fashion by computing an area under curve (*AUC*) of the coverage-specificity curve present in Fig. 8.1. Obviously, the higher the *AUC* value, the better the granular model.

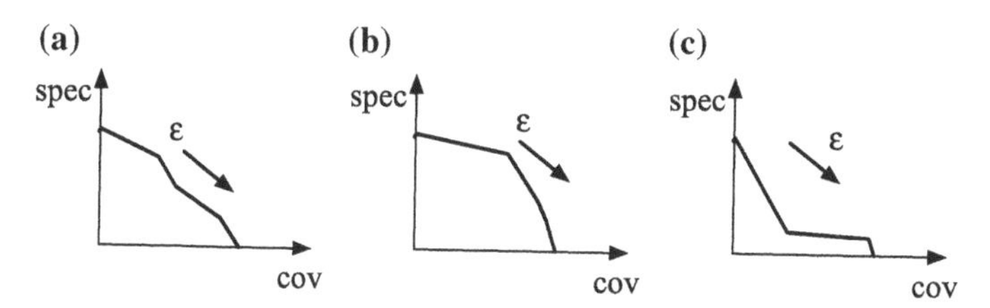

Fig. 8.2 Characteristics of coverage-specificity of granular models **a** monotonic behaviour of the relationship with the changes of ε, **b** increase of coverage and retention of specificity with the increase of ε, **c** rapid drop in specificity for increasing values of ε

Usually there are a limited number of values of e admitted in the determination of the *AUC* and then the computing of the area is expressed as the following sum:

$$\text{AUC} = \sum_{i=1}^{r} cov_i * sp_i \tag{8.7}$$

where r is the number of values of ε used in the determination of the quality of the granular model and cov_i and sp_i are the corresponding coverage and specificity values. The *AUC* value can be treated as an indicator of the global performance of the original numeric model produced when assessing granular constructs built on their basis. For instance, the quality of the original numeric models M_0 and M_0' could differ marginally but the corresponding values of their *AUC* could vary quite substantially by telling apart the models. For instance, two neural networks of quite similar topology may exhibit similar performance, however, when forming their granular generalizations, these could differ quite substantially in terms of the resulting values of the *AUC*.

As to the allocation of information granularity, the maximized coverage can be realized with regard to various alternatives as far as the data are concerned: (a) the use of the same training data as originally used in the construction of the model, (b) use the testing data and (c) usage of some auxiliary data.

8.6 Collaborative Development of Fuzzy Models

There are situations where a general model has to be built on a basis of several locally available data. Those data are reflective of the same phenomenon (system) being observed or perceived from different points of view. The input variables (features) could also be different. Such situations occur when dealing with records of patients treated at different medical facilities, customers using services of several financial institutions, etc. At the same time, the data cannot be shared because of some technical or legal constraints. A construction of a global model is then realized in a distributed mode. A way of constructing the individual models and the global can be referred to as collaborative modelling. In particular, as clustering plays a pivotal role in fuzzy rule-based models, its usage here is referred to as collaborative clustering.

8.6.1 Collaborative Clustering

In collaborative clustering, we are concerned with a collaborative development of information granules by simultaneously taking into consideration several (p) data sets. There are three possibilities to be considered as illustrated in Fig. 8.3. The two coordinates when characterizing the data are the data points and the features

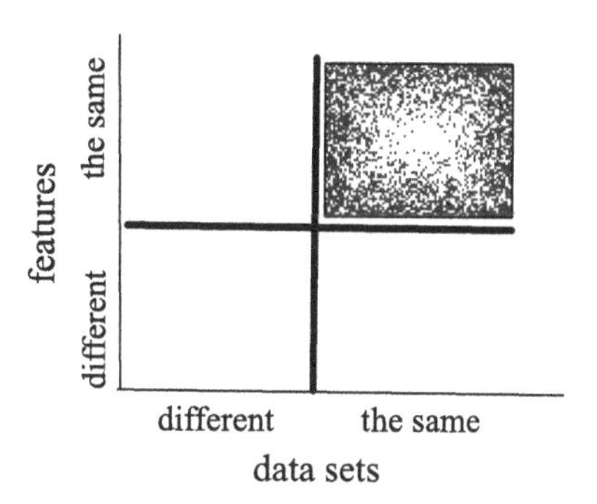

Fig. 8.3 Relationships between data sets: a perspective of data and feature spaces present in individual locally available data

(variables). As a matter of fact, the fourth one that is visualized there refers to the same data sets and this alternative is of no interest in the context of collaborative clustering.

Depending upon the relationships holding between the data involved in the collaboration process, the objective functions used in clustering are formulated accordingly. To deliver a systematic description, we adhere to the following notation. The data sets are described by $\boldsymbol{D}_1, \boldsymbol{D}_2, \ldots, \boldsymbol{D}_p$ while the corresponding feature sets associated with these data are expressed as $\boldsymbol{F}_1, \boldsymbol{F}_2, \ldots, \boldsymbol{F}_p$. The indexes in square brackets refer to the data set, say $\boldsymbol{x}_k[ii]$ is the kth data coming from the iith data set. $u_{ik}[ii]$ and $\boldsymbol{v}_i[ii]$ concern the ikth element of the partition matrix and the ith prototype in $\boldsymbol{D}_{ii}$. In the scenarios discussed below, we assume that the number of clusters in all data sets is the same and equal to c.

Different feature spaces and the same data The objective function to be minimized through clustering realized for $\boldsymbol{D}_{ii}$ comes in the following form:

$$Q_{ii} = \sum_{i=1}^{c} \sum_{k=1}^{N} u_{ik}^2[ii] \|\boldsymbol{x}_k[ii] - \boldsymbol{v}_i[ii]\|^2 \tag{8.8}$$

Here $\lambda_{ii,jj}$ is a weight expressing an extent to which the results obtained for D_{jj} influence the determination of the structure in D_{ii}. The higher the value of this index is, the stronger the impact of the structure in $\boldsymbol{D}_{jj}$ in the formation of structural dependencies in $\boldsymbol{D}_{ii}$. If $\lambda_{ii,jj}$ is set to zero, no influence is present. Observe that as we are concerned with the same data but different features, the communication mechanism is achieved by exchanging local findings expressed in terms of the partition matrices.

Different data and the same feature spaces When the data sets share the same data expressed in different feature spaces, a way to communicate (exchange) findings is to use prototypes. The objective function comes in the form

$$Q_{ii} = \sum_{i=1}^{c} \sum_{k=1}^{N} u_{ik}^2[ii] \|\boldsymbol{x}_k[ii] - \boldsymbol{v}_i[ii]\|^2 \tag{8.9}$$

Here $\mu_{ii,jj}$ are the corresponding weights establishing a way of collaboration among the data when building the clusters.

Different data and different feature spaces Collaborative clustering can be realized here assuming that there is a nonempty intersection between the data or feature spaces present in the data sets involved in collaboration. Considering that at least one of these intersections is nonempty, say there is jj such that $\boldsymbol{D}_{ii} \cap \boldsymbol{D}_{jj} \neq \emptyset$ and $\boldsymbol{F}_{ii} \cap \boldsymbol{F}_{jj} \neq \emptyset$ we have

$$Q_{ii} = \sum_{i=1}^{c}\sum_{k=1}^{N} u_{ik}^2[ii]\,||\boldsymbol{x}_k[ii] - \boldsymbol{v}_i[ii]||^2 \tag{8.10}$$

If the number of clusters varies from one data set to another, the communication is expressed in terms of the proximity matrices, Prox($U[ii]$) and Prox($U[jj]$) where Prox (U) is an $N \times N$ matrix with the entries computed as the sum of the minima of the partition matrices. The (k, l)th entry of the proximity matrix, $\mathrm{Prox}_{kl}[ii]$ reads as $\sum_{i=1}^{c} \min(u_{ik}, u_{il})$

$$Q_{ii} = \sum_{i=1}^{c[ii]}\sum_{k=1}^{N} u_{ik}^2[ii]\,||\boldsymbol{x}_k[ii] - \boldsymbol{v}_i[ii]||^2 \tag{8.11}$$

8.6.2 Fuzzy Models with Collaborative Clustering: Development Scenarios

Several situations can be distinguished when designing fuzzy models; they can be referred to as active modes of model development. To be consistent with the previous notation, the fuzzy models built on a basis of D_1, $\boldsymbol{D}_2$, $\boldsymbol{D}_p$ come in the form of the following rules:

For data set $\boldsymbol{D}_{ii}$

$$\text{if } \boldsymbol{x} \text{ is } A_i[ii] \text{ then } y \text{ is } f_i(\boldsymbol{x}, \mathbf{a}_i[ii]) \tag{8.12}$$

$i = 1, 2, \ldots, c$. The indexes in the square brackets relate to the iith data set. The output of the rule-based model resulting from $\boldsymbol{X}_{ii}$ reads as follows:

$$y[p] = \sum_{i=1}^{c} A_i[ii](\boldsymbol{x}) f_i(\boldsymbol{x}, \boldsymbol{a}_i[ii]) \tag{8.13}$$

(i) Collaborative clustering in the building of information granules in the condition parts of the rules for p fuzzy models. This way of forming granules has been discussed in the previous section. Then for each local model, one forms

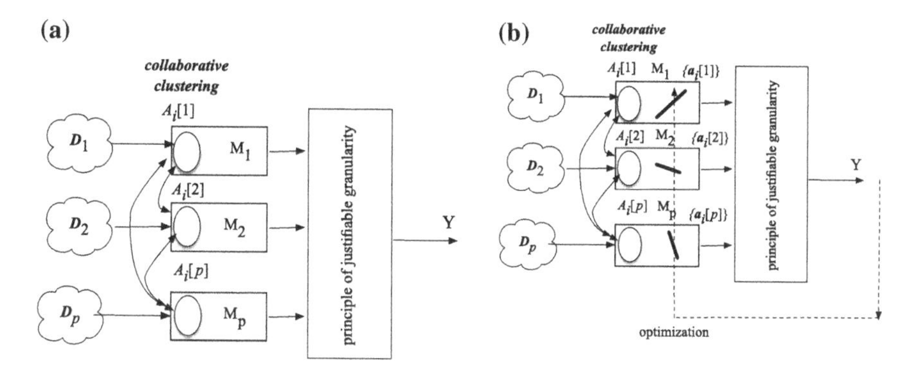

Fig. 8.4 Design of granular fuzzy models **a** use of collaborative clustering and **b** refinement of the parameters of the linear models

the conclusion part (estimation of parameters of linear functions). Then the models are arranged together in the sense their results are aggregated with the use of the principle of justifiable granularity, see Fig. 8.4a.

(ii) Collaborative clustering in the construction of the condition parts of the rules followed by the refinement of the parameters of the conclusions of the rules of the local models. Their adjustments are driven by the global performance index determined on the basis of the AUC computed for the granular results of aggregation, see Fig. 8.4b. By sweeping through several values of ξ in (8.5) the results in terms of coverage and specificity are determined and then the AUC value is obtained. The adjustments of the parameters of the local models can be completed with the use of population-based optimization methods.

For reference, one can consider a global model constructed in a passive mode (the modes discussed above can be referred to as active ones). Fuzzy models are built independently from each other and their results are combined using the principle of justifiable granularity, see Fig. 8.5. The approach is passive in the sense there is no interaction when building the fuzzy models and the aggregation is done *post mortem* once the individual models have been constructed.

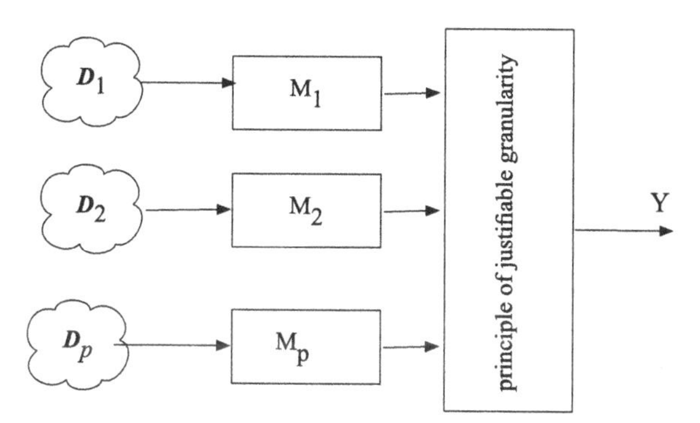

Fig. 8.5 Aggregation of local fuzzy models: a passive approach

In all situations discussed above the resulting model is granular to capture the diversity of the individual sources of knowledge (local models).

The evaluation of the quality of the global models formed as discussed above is realized by computing their *AUC* values.

8.7 Transfer Knowledge in Fuzzy Models

Transfer learning (or better to say, a transfer of knowledge) [14] is about adjusting the existing model (source of knowledge) so that it can function efficiently in a new environment. More specifically, if the model has been developed on a basis of some experimental data $\boldsymbol{D}$ (and these data were characterized by some statistical properties such as prior and conditional probabilities), it is assumed that in a new environment generating data $\boldsymbol{F}$ the knowledge conveyed by the previous model can be effectively used to update the model to the current environment rather than constructing the model from scratch.

The new data $\boldsymbol{H}$ (whose cardinality is lower than the one using which the model has been constructed) is expressed in the form of the input–output pairs $(\boldsymbol{x}_k, g_k)$, $k = 1, 2,\ldots, M$, $M \ll N$. Considering the original rule-based model described by (1), the transfer knowledge is concerned with retaining the original rules and calibrating (modifying) the input spaces by introducing (and optimizing) a continuous mapping for each input variable, say $\Phi_i(x_i)$, $i = 1, 2,\ldots, n$ so that the rules resulting from this transformation come in the following form:

Model $\boldsymbol{M}^*$

$$\text{if } \boldsymbol{\Phi}(\boldsymbol{x}) \text{ is } A_i \text{ then } y \text{ is } f_i(\boldsymbol{\Phi}(\boldsymbol{x}), \mathbf{a}_i) \tag{8.14}$$

where $\boldsymbol{\Phi} = [\Phi_1\ \Phi_2\ \Phi_n]$. The output of the model comes as a result of aggregation of activation of the individual rules and reads as follows:

$$y[p] = \sum_{i=1}^{c} A_i[ii](\boldsymbol{\Phi}(\boldsymbol{x}))f_i(\boldsymbol{\Phi}(\boldsymbol{x}), \boldsymbol{a}_i[ii]) \tag{8.15}$$

The optimization is realized by adjusting the parameters of $\boldsymbol{\Phi}$ so that the fuzzy model $\boldsymbol{M}^*$ approximates the data g_k, namely,

$$Q = \sum_{k=}^{M} (y_k - g_k)^2 \tag{8.16}$$

where y_k is the output of the model when its input is $\boldsymbol{x}_k$. The minimization is achieved by choosing a certain class of nonlinear mappings and then optimizing the parameters of $\Phi_1, \Phi_2,\ldots,\Phi_v$. The intent is to use these mappings to compensate for the shift (drift) of the probabilistic characteristics of the newly acquired data $\boldsymbol{H}$.

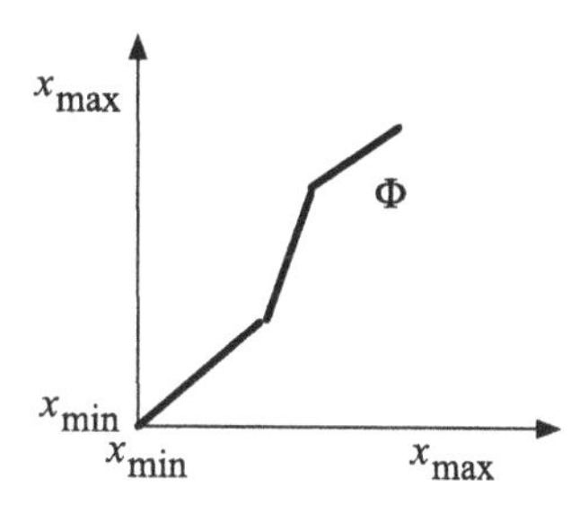

Fig. 8.6 Example piecewise linear transformation function

Some typical examples of the nonlinear mappings could be formed as piecewise linear functions, as illustrated in Fig. 8.6 or weighted combination of sigmoidal functions.

It can be observed that the nonlinear mapping transforms the original space (as being now "seen" by the individual rules) so that the changed values of the parameters of the underlying parameters of probabilities describing the data (system to be modelled) are handled by the new model M^*. In light of the nonlinearity of the mapping, some regions of the input space are contracted and some others are expanded. The essence of the resulting architecture and the transformation process is visualized in Fig. 8.7.

The mechanism of transfer knowledge can be realized in the presence of several environments producing the data $\boldsymbol{H}_1$, $\boldsymbol{H}_2$, …, $\boldsymbol{Hp}$. For each of them we form an optimal mapping Φ_1, Φ_2, …, Φ_p, see Fig. 8.8, and then realize an aggregation of the mappings into a single granular mapping $G(\boldsymbol{\Phi})$ with the use of the principle of justifiable granularity.

Owing to the existence of several environments, the granularity of the mapping is reflective of this diversity. As a direct consequence of the granularity of the

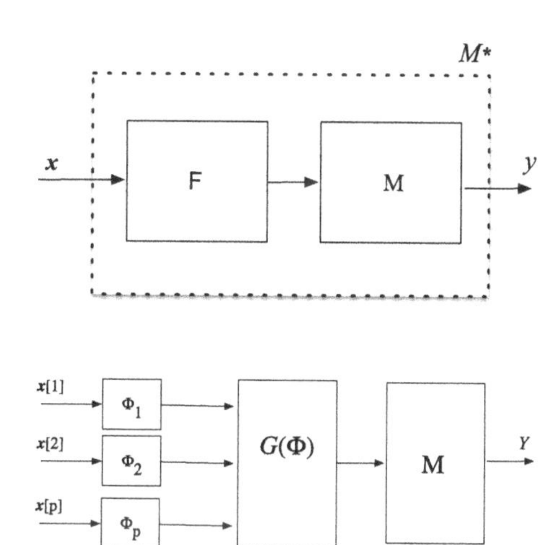

Fig. 8.7 Transfer knowledge in fuzzy rule-based models; note a nonlinear transformation of the input space

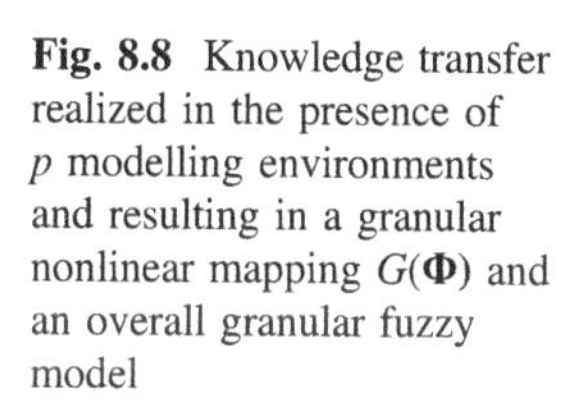

Fig. 8.8 Knowledge transfer realized in the presence of p modelling environments and resulting in a granular nonlinear mapping $G(\boldsymbol{\Phi})$ and an overall granular fuzzy model

transformation of any input carried out with the aid of $G(\Phi)$, the result of the rule-based model is granular (as a result of propagation of information granularity through the model) coming as a certain information granule Y, say, $Y = M(G(\boldsymbol{\Phi})(\boldsymbol{x}))$. The quality of the model can be evaluated in the same way as before by assessing the impact of ε on the coverage and specificity of the model developed here.

8.8 Conclusions

We offered a brief overview of fuzzy rule-based models and demonstrated that in light of new challenging modelling environments, there is a strongly motivated emergence of granular fuzzy models where the concept of information granularity and information granules of higher type/order play a pivotal role. The fundamentals of granular computing such as the principle of justifiable granularity and an optimal allocation of information granularity are instrumental in the construction of the granular models. We showed that a construction of global models on a basis of some locally available models results in an inherently granular nature of these constructs.

While the study presented here has been concentrated on fuzzy rule-based models, the developed framework exhibits a far higher level of generality and flexibility and can be applied to other modelling architectures and formalism of information granules.

Acknowledgements The support from the Canada Research Chair (CRC) and Natural Sciences and Engineering Research Council (NSERC) is gratefully acknowledged.

References

1. Alcala R, Gacto MJ, Herrera (2011) F A fast and scalable multiobjective genetic fuzzy system for linguistic fuzzy modeling in high-dimensional regression problems. IEEE Trans Fuzzy Syst. 19:666–681
2. Alonso JM, Magdalena L, Guillaume S (2006) Linguistic knowledge base simplification regarding accuracy and interpretability. Mathware Soft Comput 13:203–216
3. Bargiela A, Pedrycz W (2003) Granular computing: an introduction. Kluwer Academic Publishers, Dordrecht
5. Bezdek JC (1981) Pattern recognition with fuzzy objective function algorithms. Plenum Press, New York
4. G.E.P. Box (1976) Science and statistics. J Am Statist Assoc 71:791–799
6. Hwang C, Rhee FCH (2007) Uncertain fuzzy clustering: interval type-2 fuzzy approach to c-means. IEEE Trans Fuzzy Syst 15(12):107–120
7. Jin Y (2000) Fuzzy modeling of high-dimensional systems: complexity reduction and interpretability improvement. IEEE Trans Fuzzy Syst 8:212–221
8. Johansen TA, Babuska R (2003) Multiobjective identification of Takagi-Sugeno fuzzy models. IEEE Trans Fuzzy Syst 11:847–860

9. Mikut R, Jäkel J, Gröll L (2005) Interpretability issues in data-based learning of fuzzy systems. Fuzzy Sets Syst 150:179–197
10. Pedrycz W (2013) Granular computing: analysis and design of intelligent systems. CRC Press/Francis Taylor, Boca Raton
11. Pedrycz W, Homenda W (2013) Building the fundamentals of granular computing: a principle of justifiable granularity. Appl Soft Comput 13:4209–4218
12. Pedrycz W (2005) Knowledge-based fuzzy clustering. Wiley, New York
13. Pedrycz W, Bargiela A (2012) An optimization of allocation of information granularity in the interpretation of data structures: toward granular fuzzy clustering. IEEE Trans Syst Man Cybern Part B 42:582–590
14. Shell J, Coupland S (2015) Fuzzy transfer learning: methodology and application. Inf Sci 293:59–79
15. Yager RR (1990) Ordinal measures of specificity. Int J Gen Syst 17:57–72
16. Yao JT, Vasilakos AV, Pedrycz W (1977) Granular computing: perspectives and challenges. IEEE Trans Cybern 43(6):1977–1989
17. Zadeh LA (1997) Towards a theory of fuzzy information granulation and its centrality in human reasoning and fuzzy logic. Fuzzy Sets Syst 90:111–117
18. Zadeh LA (2005) Toward a generalized theory of uncertainty (GTU)—an outline. Inf Sci 172:1–40
19. Zhou SM, Gan JQ (2008) Low-level interpretability and high-level interpretability: a unified view of data-driven interpretable fuzzy system modelling. Fuzzy Sets Syst 159(23):3091–3131
20. Zhu B, He CZ, Liatsis P, Li XY (2012) A GMDH-based fuzzy modeling approach for constructing TS model. Fuzzy Sets Syst 189:19–29

Chapter 9
Control Law and Pseudo Neural Networks Synthesized by Evolutionary Symbolic Regression Technique

Zuzana Kominkova Oplatkova and Roman Senkerik

Abstract This research deals with synthesis of final complex expressions by means of an evolutionary symbolic regression technique—analytic programming (AP)—for novel approach to classification and system control. In the first case, classification technique—pseudo neural network is synthesized, i.e. relation between inputs and outputs created. The inspiration came from classical artificial neural networks where such a relation between inputs and outputs is based on the mathematical transfer functions and optimized numerical weights. AP will synthesize a whole expression at once. The latter case, the AP will create chaotic controller that secures the stabilization of stable state and high periodic orbit—oscillations between several values of discrete chaotic system. Both cases will produce a mathematical relation with several inputs, the latter case uses several historical values from the time series. For experimentation, Differential Evolution (DE) for the main procedure and also for meta-evolution version of analytic programming (AP) was used.

Keywords Analytic programming · Differential evolution · Control law · Pseudo neural network

9.1 Introduction

During the past five years, usage of new intelligent systems in engineering, technology, modelling, computing and simulations has attracted the attention of researchers worldwide. The most current methods are mostly based on soft computing, which is a discipline tightly bound to computers, representing a set of

---

Z.K. Oplatkova (✉) · R. Senkerik
Faculty of Applied Informatics, Tomas Bata University in Zlin,
Nam T.G. Masaryka 5555, 760 01 Zlin, Czech Republic
e-mail: oplatkova@fai.utb.cz

R. Senkerik
e-mail: senkerik@fai.utb.cz

K. Al-Begain and A. Bargiela (eds.), *Seminal Contributions to Modelling and Simulation*, Simulation Foundations, Methods and Applications,
DOI 10.1007/978-3-319-33786-9_9

methods of special algorithms, belonging to the artificial intelligence paradigm. The most popular of these methods are neural networks, evolutionary algorithms, fuzzy logic and evolutionary symbolic regression tools like genetic programming. Presently, evolutionary algorithms are known as a powerful set of tools for almost any difficult and complex optimization problem. Even the interest about the interconnection between evolutionary techniques and some other methods is spread daily.

This paper is aimed on presentation of two novel approaches of usage of evolutionary symbolic regression. In both cases, Analytic Programming is used and a complex function—relation between inputs and outputs is synthesized. The first case deals with synthesis of control law which secures the stabilization of stable state and high periodic orbit—oscillations between several values of discrete chaotic system. Second case presents the classification technique—pseudo neural networks.

Firstly, introduction into both fields is given. Afterwards, analytic programming and differential evolution are described. Design of the problem and results section for each study case follow. The conclusion finishes the chapter.

9.1.1 Control Law

The interest in the interconnection between evolutionary techniques and the control of chaotic systems is spread nowadays. First steps were done in [33, 34, 42], where the control law was based on the Pyragas method: Extended delay feedback control—ETDAS [32]. These papers were focused on how to tune several parameters inside the control technique for a chaotic system. Compared to previous research, this chapter shows a possibility of how to generate the whole control law (not only to optimize several parameters) for the purpose of stabilization of a chaotic system. The synthesis of control law is inspired by the Pyragas's delayed feedback control technique [13, 31]. Unlike the original OGY (Ott—Grebogi—York) control method [28], it can be simply considered as a targeting and stabilizing algorithm together in one package [18]. Another great advantage of the Pyragas method for evolutionary computation is the amount of accessible control parameters which can be easily tuned by means of evolutionary algorithms (EA). Apart from soft computing (artificial intelligence) methods, following methods of mathematical optimization are commonly used: the simplex method (linear programming) [21], quadratic programming [22], the branch and bound method [37] and NEH Heuristic [14]. The simplex method or linear programing is used generally for real-time and simple optimizations, mostly with unimodal cost functions, whereas the branch and bound method has proved highly successful for permutative constrained problems. NEH heuristic was developed for the optimizations of scheduling problems. Previous experiments with the connection of optimization problems and chaotic systems proved the difficulty of this task due to the highly nonlinear and erratic cost function surfaces, thus common mathematical optimization techniques could not be utilized.

In this research, instead of evolutionary algorithms (EA) utilization, analytic programming (AP) is used. The control law from the proposed system can be viewed as a symbolic structure that can be synthesized according to the requirements for the stabilization of a chaotic system. The advantage is that it is not necessary to have some "preliminary" control law and to estimate its parameters only. This system will generate the whole structure of the law even with suitable parameter values.

This research is focused on the research expansion and usage of analytic programming for the synthesis of a whole control law instead of parameters tuning for existing and commonly used control law method that is used to stabilize desired Unstable Periodic Orbits (UPO) of chaotic systems. The research presented in this chapter summarizes previous research [26, 27, 35] focused on stabilization of simple p-1 orbit—stable state, p-2 orbit and p-4 UPO—high periodic orbit (oscillations between two, respectively, four values). As an example of controlled system, Hénon map was selected for this chapter.

9.1.2 Pseudo Neural Networks

The interest about classification by means of some automatic process has been enlarged with the development of artificial neural networks (ANN). They can be used also for a lot of other possible applications like pattern recognition, prediction, control, signal filtering, approximation, etc. All artificial neural networks are based on some relation between inputs and output(s), which utilizes mathematical transfer functions and optimized weights from training process. The setting-up of layers, number of neurons in layers, estimating of suitable values of weights is a demanding procedure. On account of this fact, pseudo neural networks, which represent the novelty approach using symbolic regression with evolutionary computation, is proposed in this paper.

Symbolic regression in the context of evolutionary computation means to build a complex formula from basic operators defined by users. The basic case represents a process in which the measured data is fitted and a suitable mathematical formula is obtained in an analytical way. This process is widely known for mathematicians. They use this process when a need arises for mathematical model of unknown data, i.e. relation between input and output values. The proposed technique is similar to synthesis of analytical form of mathematical model between input and output(s) in training set used in neural networks. Therefore, it is called Pseudo Neural Networks.

Initially, John Koza proposed the idea of symbolic regression done by means of a computer in Genetic Programming (GP) [1, 16, 17]. The other approaches are e.g. Grammatical Evolution (GE) developed by Conor Ryan [23] and here described Analytic Programming [25, 39, 40].

The above-described tools were recently commonly used for synthesis of artificial neural networks but in a different manner than is presented here. One possibility is the usage of evolutionary algorithms for optimization of weights to obtain the ANN training process with a small or no training error result. Some other approaches represent the special ways of encoding the structure of the ANN either into the individuals of evolutionary algorithms or into the tools like Genetic Programming. But all of these methods are still working with the classical terminology and separation of ANN to neurons and their transfer functions [7]. In this paper, the proposed technique synthesizes the structure without a prior knowledge of transfer functions and inner potentials. It synthesizes the relation between inputs and output of training set items used in neural networks so that the items of each group are correctly classified according the rules for cost function value. The data set used for training is Iris data set (Machine Learning Repository [8, 20]). It is a very well-known benchmark data set for classification problem, which was introduced by Fisher for the first time.

9.2 Soft Computing Techniques

For both above-described problems, one soft computing technique will find the solution. The result is the same from some point of view—complex structure which has some relation between inputs and output. In the first case, historical data from time series of the activity of the chaotic system is taken as inputs. In the case of pseudo neural networks, features (attributes) of classified items are applied as inputs. The inputs are part of the synthesized mathematical expression via Analytic programming.

9.2.1 Analytic Programming

Basic principles of the AP were developed in 2001 [25, 39, 41, 43]. Until that time, genetic programming (GP) and grammatical evolution (GE) had been known. GP uses genetic algorithms while AP can be used with any evolutionary algorithm, independently on individual representation. To avoid any confusion, based on use of names according to the used algorithm, the name—Analytic Programming was selected, since AP represents synthesis of analytical solution by means of evolutionary algorithms.

The symbolic regression can be used in the different areas; for design of electronic circuits, optimal trajectory for robots design, data approximation, classical neural networks synthesis and many other applications [1, 16, 17, 23, 25, 39, 40]. Everything depends on the user-defined set of operators.

The core of AP is based on a special set of mathematical objects and operations. The set of mathematical objects is set of functions, operators and so-called terminals (as well as in GP), which are usually constants or independent variables. This set of variables is usually mixed together and consists of functions with different number of arguments. Because of a variability of the content of this set, it is called as "general functional set"—GFS. The structure of GFS is created by subsets of functions according to the number of their arguments. For example GFS_{all} is a set of all functions, operators and terminals, GFS_{3arg} is a subset containing functions with only three arguments, GFS_{0arg} represents only terminals, etc. The subset structure presence in GFS is vitally important for AP. It is used to avoid synthesis of pathological programs, i.e. programs containing functions without arguments, etc. The content of GFS is dependent only on the user. Various functions and terminals can be mixed together [25, 41, 43].

The second part of the AP core is a sequence of mathematical operations, which are used for the program synthesis. These operations are used to transform an individual of a population into a suitable program. Mathematically stated, it is a mapping from an individual domain into a program domain. This mapping consists of two main parts. The first part is called discrete set handling (DSH) (See Fig. 9.1) [19, 41] and the second one stands for security procedures which do not allow synthesizing pathological programs. The method of DSH, when used, allows handling arbitrary objects including nonnumeric objects like linguistic terms {hot, cold, dark…}, logic terms (True, False) or other user-defined functions. In the AP, DSH is used to map an individual into GFS and together with security procedures creates the above mentioned mapping which transforms arbitrary individual into a program.

AP needs some evolutionary algorithm [39] that consists of population of individuals for its run. Individuals in the population consist of integer parameters, i.e. an individual is an integer index pointing into GFS. The creation of the program can be schematically observed in Fig. 9.2.

The detailed description is represented in [24, 41, 43].

An example of the process of the final complex formula synthesis (according to the Fig. 9.2) follows.

The number 1 in the position of the first parameter means that the operator plus (+) from GFS_{all} is used (the end of the individual is far enough). Because the operator + must have at least two arguments, the next two index pointers 6 (sin from GFS) and 7 (cos from GFS) are dedicated to this operator as its arguments. The two functions, sin and cos, are one-argument functions, therefore the next

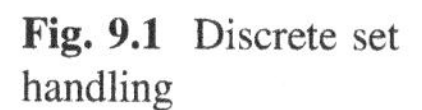
Fig. 9.1 Discrete set handling

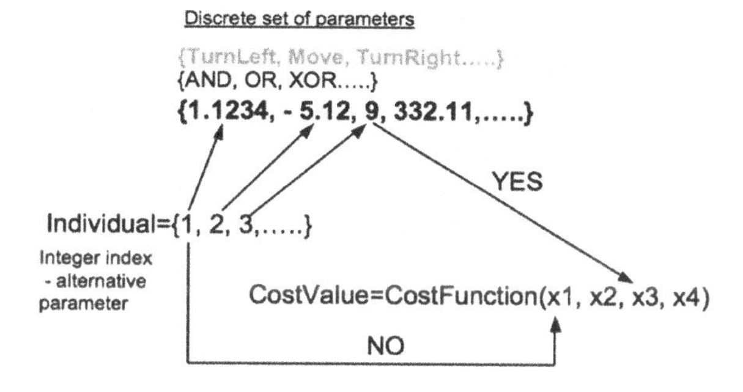

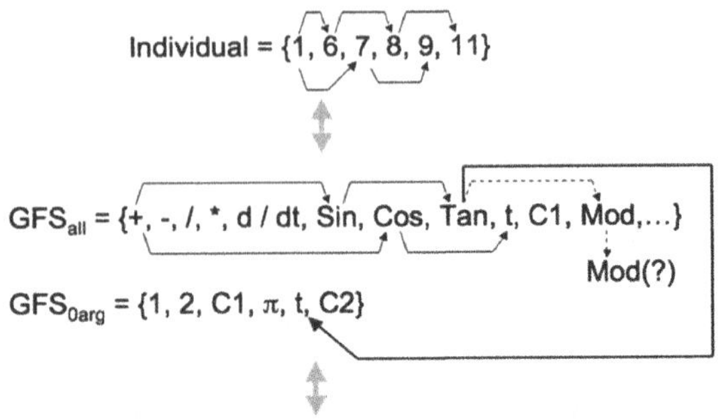

Fig. 9.2 Main principles of AP

unused pointers 8 (tan from GFS) and 9 (t from GFS) are dedicated to the sin and cos functions. As an argument of cos, the variable t is used, and this part of the resulting function is closed (t has zero arguments) in its AP development. The one-argument function tan remains, and there is one unused pointer 11, which stands for Mod in GFS_{all}. The modulo operator needs two arguments but the individual in the example has no other indices (pointers, arguments). In this case, it is necessary to employ security procedures and jump to the subset with GFS_{0arg}. The function tan is mapped on t from GFS_{0arg} which is on the 11th position, cyclically from the beginning.

AP exists in three versions—basic without constant estimation, AP_{nf}—estimation by means of nonlinear fitting package in Mathematica environment and AP_{meta}—constant estimation by means of another evolutionary algorithms; meta means metaevolution.

AP_{basic} stands for the version where constant estimation is done in the same way as in genetic programming. If for example, data approximation needs to estimate coefficients in the approximated polynomial or move the basic curve from the axes origin. In the AP_{basic} the user has to assign a set of constant values into GFS. It means a huge enlargement of the functional sets and deceleration of the evolutionary procedure. Therefore, two other strategies were adopted—AP_{nf} and AP_{meta}.

These two versions of AP use the constant K which is indexed during the evolution (9.1). When K is needed, a proper index is assigned—K_1, K_2,... K_n (9.2). Numeric values to indexed Ks are estimated (9.3) either via nonlinear fitting methods in the Mathematica environment (www.wolfram.com)—AP_{nf} or via the second evolutionary algorithm—AP_{meta}.

$$\frac{x^2 + K}{\pi^K} \tag{9.1}$$

$$\frac{x^2 + K_1}{\pi^{K_2}} \tag{9.2}$$

$$EA_{master} \Rightarrow program \Rightarrow K_{indexing} \Rightarrow EA_{slave} \Rightarrow K_{estimation} \Rightarrow final \cdot solution$$

Fig. 9.3 Schema of AP procedures

$$\frac{x^2 + 3.156}{\pi^{90.78}} \tag{9.3}$$

AP_{meta} is a time consuming process and the number of cost function evaluations, which is one of comparative factors, is usually very high. This fact is given by two evolutionary procedures (Fig. 9.3).

EA_{master} is the main evolutionary algorithm for AP, EA_{slave} is the second evolutionary algorithm inside AP. Thus, the number of cost function evaluation (CFE) is given by (9.4).

$$CFE = EA_{master} * EA_{slave} \tag{9.4}$$

9.2.2 Metaevolution

Generally, metavolution means evolution of evolution [3, 4, 12, 15, 25]. Metaevolutionary techniques for optimization tasks belong to soft computing methods as well as evolutionary algorithms [1, 25] and symbolic regression [25]. First attempts of researchers were in the usage of an evolutionary algorithm for tuning or controlling of another evolutionary technique [2, 5]. During this process, usually the best types of evolutionary operators and settings of their parameters were evolutionarily selected. Tuning of parameters is the usage of the best-found values which will be set up at the initialization of the evolutionary algorithm and used with the same values for a whole process. Compared to this, parameter control adjusts the values during the evolutionary process. It is adapted either by means of a predetermined rule or some kind of self-adaptation [36].

Another approach is to let evolution create a structure and parameters of the used evolutionary operators such as selection crossover or mutation. Diosan and Oltean use Meta Genetic Programming [3] for the evolutionary design of evolutionary algorithms. Analytic programming was also used in Oplatkova [25] with symbolic regression principles to breed a completely new structure of the optimization algorithm of evolutionary character which comes from the basic selection of operators in the AP settings.

The last technique of metaevolution is discussed in the presented applications in this chapter. This is the estimation of coefficients in symbolic regression when two evolutionary algorithms help each other. One evolutionary algorithm drives the main process of symbolic regression, in this case analytic programming, and the second is used for the constant estimation. This meta approach of analytic programming has to be used when the constants are not possible to estimate in another manner because of the character of the problem. In data approximation tasks, there can be used a technique from nonlinear fitting package, which is adopted in Wolfram Mathematica environment, because the problem is designed so that the found constants (e.g. coefficients of polynomials) move the basic shape of the curve around the coordinate system. It is not possible to employ such a package in the case of the synthesis of control laws for chaotic systems or Pseudo ANN. These applications do not use the found result as a model which could be adjusted to some "measured" values in the sense of interpolation but the found solution is used further as a part of complex technique to find a quality of the solution and cost function estimation.

9.2.3 *Differential Evolution*

As mentioned above, the Analytic Programming needs an evolutionary algorithm for the optimization—finding the best shape of the complex formula. This research used one evolutionary algorithm Differential Evolution [29] for both parts—main procedure and metaevolutionary part of the analytic programming.

DE is a population-based optimization method that works on real-number-coded individuals [29]. For each individual $\vec{x}_{i,G}$ in the current generation G, DE generates a new trial individual $\vec{x}'_{i,G}$ by adding the weighted difference between two randomly selected individuals $\vec{x}_{r1,G}$ and $\vec{x}_{r2,G}$ to a randomly selected third individual $\vec{x}_{r3,G}$. The resulting individual $\vec{x}'_{i,G}$ is crossed-over with the original individual $\vec{x}_{i,G}$. The fitness of the resulting individual, referred to as a perturbed vector $\vec{u}_{i,G+1}$, is then compared with the fitness of $\vec{x}_{i,G}$. If the fitness of $\vec{u}_{i,G+1}$ is greater than the fitness of $\vec{x}_{i,G}$, then $\vec{x}_{i,G}$ is replaced with $\vec{u}_{i,G+1}$; otherwise, $\vec{x}_{i,G}$ remains in the population as $\vec{x}_{i,G+1}$. DE is quite robust, fast and effective, with global optimization ability. It does not require the objective function to be differentiable, and it works well even with noisy and time-dependent objective functions. Description of used DERand1Bin strategy is presented in (9.5). Please refer to [29, 30] for the description of all other strategies.

$$u_{i,G+1} = x_{r1,G} + F \bullet \left(x_{r2,G} - x_{r3,G}\right) \tag{9.5}$$

9.2.4 *Artificial Neural Networks*

Artificial neural networks are inspired in the biological neural nets and are used for complex and difficult tasks [6, 9, 10, 38]. The most often usage is classification of objects as also in this case. ANNs are capable of generalization and hence the classification is natural for them. Some other possibilities are in pattern recognition, control, filtering of signals and also data approximation and others.

There are several kinds of ANN. Simulatiohave a value betwens were based on similarity with feedforward net with supervision. ANN needs a training set of known solutions to be learned on them. Supervised ANN has to have input and also required output.

The neural network works so that suitable inputs in numbers have to be given on the input vector. These inputs are multiplied by weights which are adjusted during the training. In the neuron, the sum of inputs multiplied by weights is transferred through mathematical function like sigmoid, linear, hyperbolic tangent, etc. Therefore, ANN can be used for data approximation [10]—a regression model on measured data, relation between input and required (measured data) output.

These single neuron units (Fig. 9.4) are connected to different structures to obtain different structures of ANN (e.g. Figs. 9.5 and 9.6), where $\sum \delta = TF[\sum (w_i x_i + b w_b)]$ and $\sum = TF[\sum (w_i x_i + b w_b)]$; TF is logistic sigmoid function in this case.

The example of relation between inputs and output can be shown as a mathematical form (9.6). It represents the case of only one neuron and logistic sigmoid function as a transfer function.

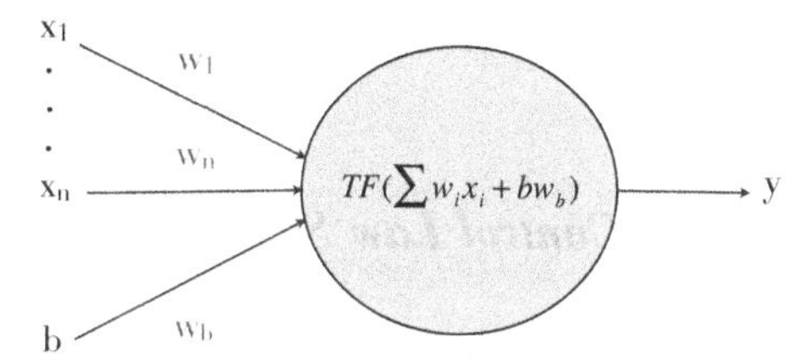

Fig. 9.4 Neuron model, where TF (transfer function like sigmoid), x_1–x_n (inputs to neural network), b—bias (usually equal to 1), w_1—w_n, w_b—weights, y—output

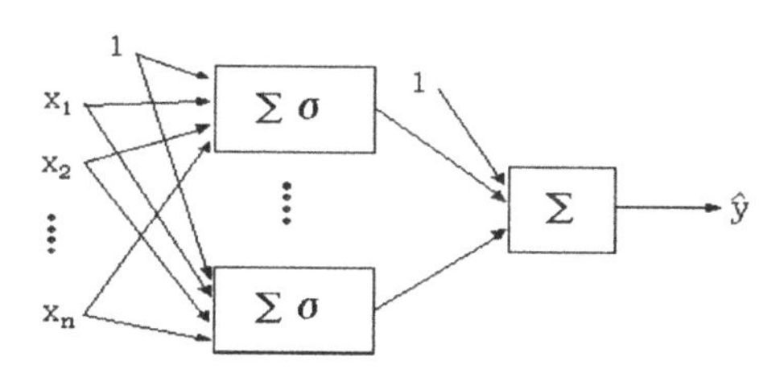

Fig. 9.5 ANN models with one hidden layer

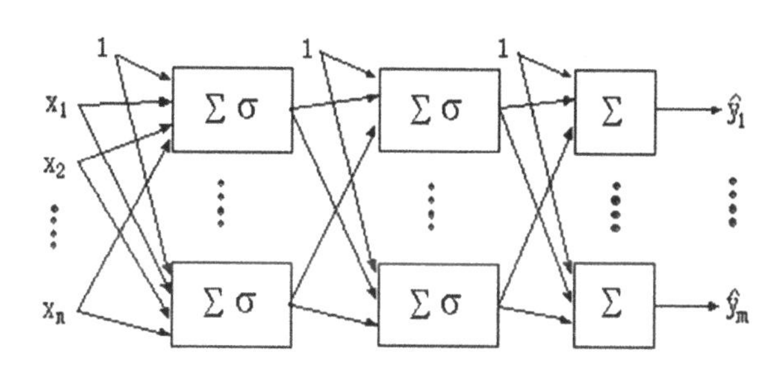

Fig. 9.6 ANN models with two hidden layers and more outputs

$$y = \frac{1}{1 + e^{-(x_1 w_1 + x_2 w_2)}}, \tag{9.6}$$

where

y output

x_1,x_2 inputs

w_1,w_2 weights

The aim of the proposed technique is to find similar relation to (9.6). This relation is completely synthesized by evolutionary symbolic regression—Analytic Programming.

9.3 Problem Design

Both presented applications used Analytic programming for a synthesis of a mathematical complex formula where several inputs are used for calculation. In the case of control law synthesis, historical data from the chaotic system time-dependent behaviour are calculated as inputs into the mathematical expression. More details are given in the following section.

In the case of pseudo neural networks, inputs stands for features or attributes of classified items as described below in detail.

9.3.1 Control Law Synthesis

9.3.1.1 Used Chaotic System

The chosen example of chaotic system was the two dimensional Hénon map in form (9.7):

$$\begin{aligned} x_{n+1} &= a - x_n^2 + by_n \\ y_{n+1} &= x_n \end{aligned} \tag{9.7}$$

This is a model invented with a mathematical motivation to investigate chaos. The Hénon map is a discrete-time dynamical system, which was introduced as a simplified model of the Poincaré map for the Lorenz system. It is one of the most studied examples of dynamical systems that exhibit chaotic behaviour. The map depends on two parameters, a and b, which for the canonical Hénon map have values of a = 1.4 and b = 0.3. For these canonical values the Hénon map is chaotic

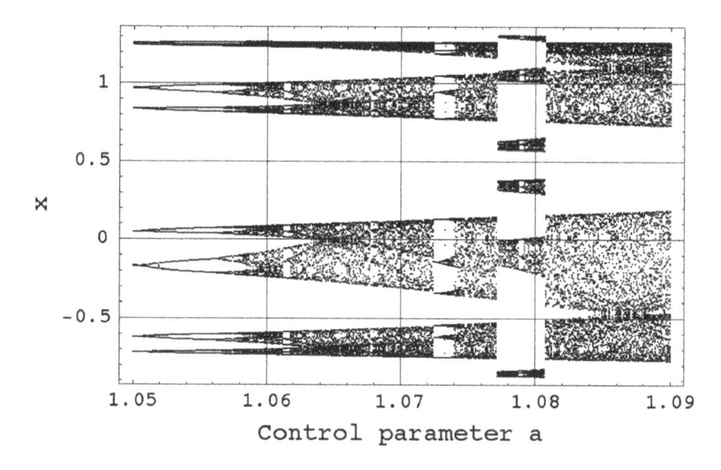

Fig. 9.7 Bifurcation diagram of Hénon Map

Table 9.1 Values for UPOs for Hénon map

UPO	Values for unperturbed system
1-p (stable state)	$x_F = 0.8$
2-p (oscillation between 2 points)	$x_F = -0.56$ and 1.26
4-p (oscillation between 4 points)	$x_F = 0.139$, 1.4495, −0.8594 and 0.8962

[11]. The example of this chaotic behaviour can be clearly seen from bifurcation diagram—Fig. 9.7.

Figure 9.7 shows the bifurcation diagram for the Hénon map created by plotting of a variable x as a function of the one control parameter for the fixed second parameter.

Values for different UPOs are given in Table 9.1. The aim of the stabilization is to reach these values.

9.3.1.2 Problem and Cost Function Design

The methodology used in these simulations is divided into two parts—control laws for p-1 orbit (stable state) and higher periodic orbits p-2 and p-4 UPO (oscillation between two, respectively, four values). In the first case, the inspiration for preparation of sets of basic functions and operators for AP was the simpler TDAS control method (9.8) and its discrete form given in (9.9).

$$F(t) = K[x(t-\tau) - x(t)] \tag{9.8}$$

$$F_n = K(x_{n-m} - x_n) \quad (9.9)$$

This means that only current output value x_n and one previous x_{n-1} were used in the set of basic functions together with constants, operators such as plus, minus and power.

The latter case was inspired by the method ETDAS due to the recursive attributes of the delay equation S utilizing previous states of the system. Therefore, the data set for AP was expanded and covers a longer system output history.

The original control method—ETDAS has the form (9.10).

$$F(t) = K[(1 - R)S(t - \tau_d) - x(t)]$$
$$S(t) = x(t) + RS(t - \tau_d) \quad (9.10)$$

where: K and R are adjustable constants, F is the perturbation; S is given by the delay equation utilizing previous states of the system and τ_d is a time delay. The original control method—ETDAS in the discrete form has the form (9.11).

$$F_n = K[(1 - R)S_{n-m} - x_n]$$
$$S_n = x_n + RS_{n-m} \quad (9.11)$$

where: m is the period of m-periodic orbit to be stabilized. The perturbation F_n in Eq. (9.11) may have arbitrarily large value which can cause the diverging of the system. Therefore, F_n should have a value between $-F_{max}, F_{max}$. In this chapter, a suitable F_{max} value was taken from the previous research [34].

The proposal for the cost function comes from the simplest Cost Function (CF) which is used within Analytic programming and different applications. The idea was to minimize the area created by the difference between the required state and the real system output on the whole simulation interval—τ_i.

Because of the stabilization of an extremely sensitive chaotic system, another universal cost function with the possibility of adding penalization rules had to be used. It was synthesized from the simple CF and other terms were added. In this case, it is not possible to use the simple rule of minimizing the area created by the difference between the required and actual state on the whole simulation interval —τ_i, due to many serious reasons, for example: including of initial chaotic transient into the final CF value or degrading of the possible best solution by phase shift of higher periodic orbit, which represents the oscillations between several values.

This CF is in general based on searching for a desired stabilized periodic orbit and thereafter calculation of the difference between the desired and found actual periodic orbit on the short time interval—τ_s (20 iterations) from the point where the first minimal value of the difference between the desired and actual system output is found. Such a design of CF should secure the successful stabilization of either p-1 orbit (stable state) or a higher periodic orbit anywise phase shifted. The CF_{Basic} has the form (9.12).

$$CF_{Basic} = pen_1 + \sum_{t=\tau 1}^{\tau 2} |TS_t - AS_t| \tag{9.12}$$

where:

TS—target state, AS—actual state

τ_1—the first min value of the difference between TS and AS, τ_2—the end of the optimization interval ($\tau_1 + \tau_s$)

$pen_1 = 0$ if $\tau_i - \tau_2 \geq \tau_s$; $pen_1 = 10*(\tau_i - \tau_2)$ if $\tau_i - \tau_2 < \tau_s$ (i.e. late stabilization).

9.3.1.3 Settings

The novelty of the metaevolutionary approach represents the synthesis of a feedback control law F_n (perturbation) inspired by the original ETDAS or TDAS control method. The perturbation is the feedback to the system which helps to stabilize it.

Therefore the basic set of elementary functions for AP was selected as follows. The main items are several previous values (data) to current value in the controlled chaotic system.

GFS_{2arg} = +, −, /, *, ^
GFS_{0arg} = $data_{n-1}$ to $data_n$, K (for p-1 orbit)
GFS_{0arg} = $data_{n-9}$ to $data_n$, K (for p-2 orbit).
GFS_{0arg} = $data_{n-11}$ to $data_n$, K (for p-4 orbit).

Settings of EA parameters for both processes in AP were based on performed numerous experiments and simulations with AP_{meta} is given in Tables 9.2 and 9.3; DE was used for main AP process and also in the second evolutionary process.

Table 9.2 DE settings for main process

PopSize	20
F	0.8
CR	0.8
Generations	50
Max. CF Evaluations (CFE)	1000

Table 9.3 DE settings for metaevolution

PopSize	40
F	0.8
CR	0.8
Generations	150
Max. CF Evaluations (CFE)	6000

Table 9.4 Simulation results for 1-p, 2-p and 4-p UPOs

Law	Control law with coefficients	CF Value	UPO value of stabilized system	Figure
1-p UPO—stable state				
1	$F_n = \frac{(x_{n-1} - x_n - 1.62925)(x_{n-1}x_n - x_n^2)}{2x_{n-1}}$	1.3323×10^{-15}	0.8	Figure 9.8a
2	$F_n = -\frac{0.781971(x_{n-1}x_n - x_n^2)}{x_{n-1}}$	1.3323×10^{-15}	0.8	Figure 9.8b
2-p UPO—oscillation between 2 points				
1	$F_n = 0.523744(x_{n-1} + x_n - 0.700001)$	1.39845×10^{-5}	−0.56–1.26	Figure 9.8c
2	$F_n = 0.20375x_{n-8}(x_{n-4} - x_{n-3} + 1.87967)x_{n-3}(x_{n-2} - x_n)$	1.70211×10^{-5}	−0.56–1.26	Figure 9.8d
4-p UPO—oscillation between 4 points				
1	$F_n = -0.527409x_{n-8}x_{n-7}x_{n-3}(x_{n-4} - x_n)$	0.0984	0.13, 1.45, −0.86, 0.89	Figure 9.8e
2	$F_n = \frac{x_{n-7}x_{n-6}(10.0667 + x_n)}{59.4863 + \frac{-0.0174x_{n-6} - 52.191}{\frac{x_n 5.9706}{x_{n-3}} + 39.4742 + x_{n-2}}}$	0.7095	0.13, 1.45, −0.86, 0.89	Figure 9.8f

9.3.1.4 Results

The examples of new synthesized feedback control laws F_n (perturbation) for the controlled Hénon map, for stabilization of 1-p, 2-p and 4-p UPOs, are given in Table 9.4. Simulation outputs are depicted in Fig. 9.8. The minimal final CF value is very close to zero for all selected examples. This gives weight to the argument, that AP is able to synthesize various types of control laws, securing the precise and fast stabilization with machine numerical precision on the unstable periodic orbit of a real chaotic system. The only difference across all simulation results was the speed of stabilization. Nevertheless this quality parameter was not included in the CF this time.

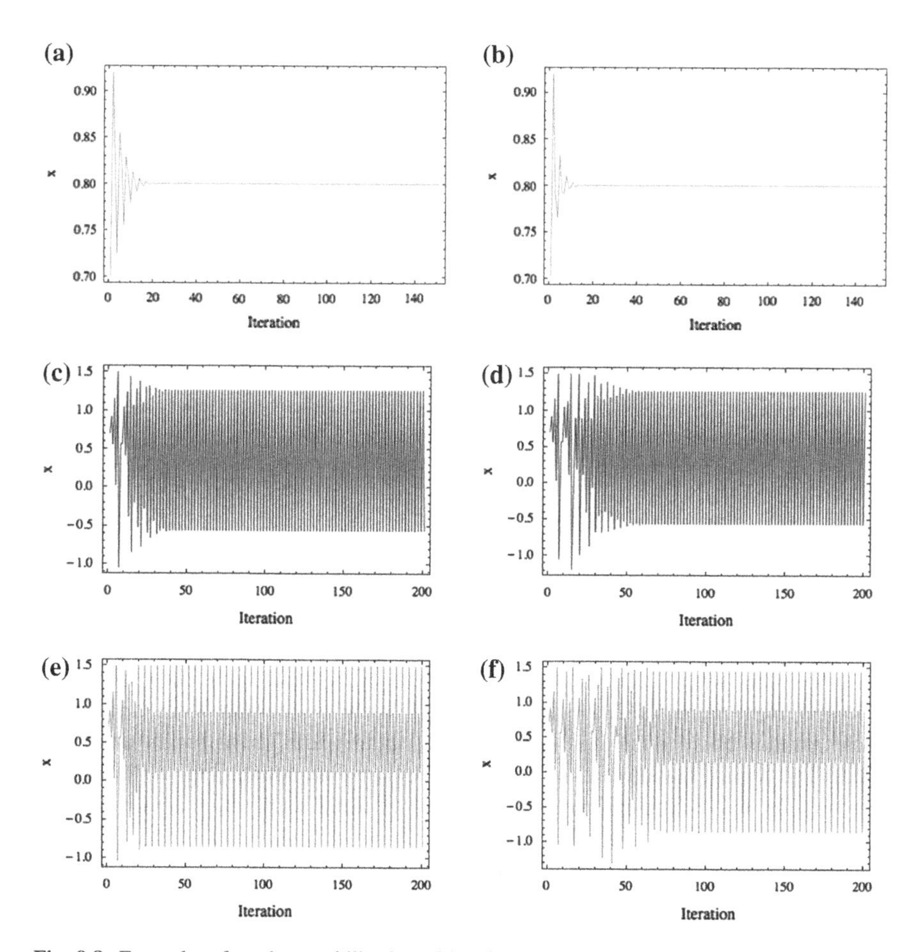

Fig. 9.8 Examples of results—stabilization of 1-p, 2-p and 4-p UPOs for Hénon map by means of control laws given in Table 9.4

9.3.1.5 Discussion

Obtained results show that synthesized control laws provided better results than the original control method which served as an inspiration.

Presented data and statistical comparison can be summarized as follows:

All simulations were performed at least 50x to obtain statistics. All cases have found the solution for the stabilization of chaotic systems. The number of cost function evaluations for 32 millions per one simulation means that the consumed time is really high. The future research will be supposed to search for time efficiency and decreasing of the simulation time.

Also the cost function is important for the successful process. The simple evolutionary approach is easy to implement, it is very fast and gives satisfactory results. But the quality of results is restricted by the limitations of the mathematical formulas, control laws, models, etc., for which the parameters are tuned by EA.

9.3.2 Classification Case

Evolutionary symbolic regression can be used also for classification applications. It depends on the user's view how the mathematical complex functional can be interpreted. If the attributes (features) of the classification items are used as inputs into the mathematical expression, then the relation can be viewed as a relation between inputs and output(s), i.e. inputs are transferred to the output via mathematical expression. If the trained classical neural network is analyzed and the relation between inputs and outputs is extracted, it is also based on mathematical complex function.

9.3.2.1 Iris Plant Data Set Definition

For this classification problem, iris data set was used (Machine Learning Repository [8, 20]). This set contains 150 instances. Half amount was used as training data and the second half was used as testing data. The data set contains 3 classes of 50 instances each, where each class refers to a type of iris plant. One class is linearly separable from the other 2; the latter are NOT linearly separable from each other. Each instance has 4 attributes (sepal length, sepal width, petal length and petal width) and type of class—iris virginica (Fig. 9.9), iris versicolor (Fig. 9.10) and iris sentosa (Fig. 9.11). The attributes were of real values. Usually, the class is defined by 3 output neurons in classical artificial neural net and binary code. In this chapter, two study apporaches with AP were used:

(a) just one output—MISO (multiple inputs single output) system where three classes were designed as continues value of one output, i.e. iris setosa was -1 for output less than -1, iris virginica has value 1 for output greater than 1 and iris versicolor was 0 for outputs between -1 and 1;

Fig. 9.9 Iris virginica

Fig. 9.10 Iris versicolor

Fig. 9.11 Iris setosa

(b) AP was used more times to create more independent outputs—MIMO (multiple inputs multiple outputs) system. Their combination gives the result of the classification. The inspiration came from the classical artificial neural networks where the class is defined by 3 output nodes and binary code. After the ANN training, it can be separated an equation for each output node—relation between inputs, weights and the output node. The training is done in a parallel way—all weights, which are set up produce the final equations in "one step". AP was able to create pseudo neural net structure between inputs and also three outputs. The synthesis of each relation between inputs and outputs was done in serial and independent way. It is different from the previous approach where only one output node was used and continuous classification within defined intervals of output values for each class. Both approaches seem to be useful. Within more classes, more output nodes are needed. The continuous classification will not be able to cover e.g. 20 classes. There will be very small differences between intervals and it is not usable. The output values were designed as it is usual with ANN in this case. The combination of zeros and ones, i.e. iris setosa was (1,0,0), iris virginica (0,1,0) and iris versicolor (0,0,1). The training patterns were setup as the appropriate item in the correct class to 1 and the other two kinds of plants were zero and this was cyclically done for all cases.

The cost function value is given as the basic one, in Eq. (9.13) for both cases, i.e. if the cv is equal to zero, all training patterns are classified correctly.

$$cv = \sum_{i=1}^{n} |required\ Output - current\ Output| \tag{9.13}$$

9.3.2.2 Results

This study case used the same settings for Differential evolution as the study case with control law synthesis (Tables 9.2 and 9.3). Metaevolutionary approach is needed to find optimal values of constants in the structure of pseudo neural networks.

The set of elementary functions for AP was inspired in the items used in classical artificial neural nets. It is also similar to control law synthesis. The components of input vector x contain values of each attribute (x_1, x_2, x_3, x_4).

Basic set of elementary functions for AP:

GFS2arg = + , −, /, *, ^, exp
GFS0arg = x_1, x_2, x_3, x_4, K

Total number of cost function evaluations for AP was 1000, for the second EA it was 6000, together 6 millions per each simulation.

From carried simulations, several notations of input dependency were obtained, e.g. (9.14) and (9.15) for MISO system and continues classification. The case (9.14) had a training error equal to 2 misclassified items (it means 2.66 % error, 97.34 % success from all 75 training patterns) and testing error equal to 3 misclassified items (4 % error, 96 % success). The case (9.15) was worse, during training phase the error equal to 3 misclassified patterns was reached and testing error was equal to 5.

Following numbers (9.16) are the outputs from the pseudo neural network for the training set, which has been then saturated to 3 numbers −1, 0 and 1 according to the class type interval (9.17). The (9.18) and (9.19) contain numbers for testing set obtained is the same way as the training set.

$$y = \left(-\frac{7.57336\, e^{-x_3} x_2}{x_4} - x_3 + x_4\right) \exp\left(-6.91605 \times 10^{117^{344.416 e^{-x_2^{x_4}} (x_1 + 1.0796\times 10^{-233 x_1 - e^{x_2}})_{-e^{-x_4 - e^{x_2}}}}}\right) \tag{9.14}$$

$$y = {}^{(e^{-0.0066007(x_4+693.44)})^{-0.00311885(x_3+e^{x_3}-144.991)} - 953.535\, e^{x_3(-x_4-637.961)}} \left(-x_2 - x_4 + \frac{6.11200456962 \times 10^{625}}{(x_2 x_4)^{223.656}}\right) \tag{9.15}$$

$$\begin{aligned} \text{classical output} = \{&-2.32401, -1.82401, -2.02399, -1.92404, -2.42401, -2.51714, -2.1247, -2.22404, \\ &-1.72401, -2.01559, -2.52404, -2.22406, -1.91558, -1.91555, -2.82397, -3.01703, \\ &-2.51692, -2.2247, -2.52481, -2.52474, -2.22408, -2.31703, -2.42392, -1.80036, \\ &-2.22413, -0.100247, 0.162017, 0.259989, 0.561469, 0.56151, 0.0595996, 0.344323, \\ &-0.153875, -0.0407742, 0.403247, 0.245674, 0.36354, 0.0445453, 0.199753, \\ &-0.037034, 0.00106313, 0.362017, -0.45568, 1.16202, -0.0659825, 1.08403, 0.0614693, 0.859989, \\ &-0.163289, -0.0396526, 4.23024, 1.94265, 2.4494, 1.37783, 2.90987, 2.44134, 1.45294, 1.37241, \\ &1.77628, 3.92834, 1.8341, 1.94088, 2.45401, 2.53511, 4.15996, 3.21541, 1.27861, 2.0981, \\ &3.79167, 1.15948, 3.20947, 2.23613, 2.21791, 1.58326, 2.15171\} \end{aligned} \tag{9.16}$$

$$\begin{aligned} \text{saturated output to class type} = \{&-1, -1, -1, -1, -1, -1, -1, -1, -1, -1, -1, -1, -1, \\ &-1, -1, -1, -1, -1, -1, -1, -1, -1, -1, -1, -1, 0, \\ &0, 0, 0, 0, 0, 0, 0, 0, 0, 0, 0, 0, 0, 0, 0, 0, \\ &1., 0, 1., 0, 0, 0, 0, 1., 1., 1., 1., 1., 1., 1., 1., 1., \\ &1., 1., 1., 1., 1., 1., 1., 1., 1., 1., 1., 1., 1., 1., 1., 1.\} \end{aligned} \tag{9.17}$$

$$\begin{aligned} \text{testing classical output} = \{&-1.82406, -2.01709, -2.32404, -2.22401, -2.02406, -1.92406, -2.01703, \\ &-3.01559, -3.02401, -2.01559, -2.02397, -2.32399, -2.01559, -1.82399, \\ &-2.22404, -2.22466, -1.02466, -2.02399, -1.87359, -2.41725, -1.7247, -2.62406, -2.02401, \\ &-2.52404, -2.12401, 0.101063, 0.299317, 0.949556, 0.462017, -0.354326, 0.0342867, \\ &-0.154778, -0.0607432, 0.941977, 0.362017, 0.245497, 0.261003, 0.559973, \\ &-0.138905, 0.361469, 0.0376652, 0.10019, 0.0389384, -0.053875, 0.160721, -0.361698, \\ &-0.0392787, -0.0396526, -0.063559, 0.0610953, 1.07473, 1.48403, 1.28326, 2.65286, 0.637876, 1.83378, \\ &1.22094, 3.11249, 0.558975, 0.495828, 3.40353, 3.55155, 1.17861, 1.28403, 2.35517, \\ &3.85155, 3.31839, 1.94265, 3.2065, 4.23593, 3.4169, 2.14354, 2.03309, 3.01393, 1.28171 \end{aligned} \tag{9.18}$$

$$\begin{aligned} \text{testing saturated output to class type} = \{&-1, -1, -1, -1, -1, -1, -1, -1, -1, -1, -1, -1, -1, -1, -1, -1, -1, -1, \\ &-1, -1, -1, -1, -1, -1, -1, 0, 0, 0, 0, 0, 0, 0, 0, 0, 0, 0, 0, 0, 0, 0, 0, 0, 0, 0, 0, 0, 0, 0 \\ &0, 0, 1., 1., 1., 1., 0, 1., 1., 1., 0, 0, 1., 1., 1., 1., 1., 1., 1., 1., 1., 1., 1., 1., 1., 1., 1.\} \end{aligned} \tag{9.19}$$

From the above values, it can be visible which items were classified in a wrong group. During training iris versicolor was misclassified twice as iris sentosa, in the testing phase it was viceversa—iris sentosa as iris versicolor.

For MIMO system, the following Eqs. (9.20)–(9.22) are just examples of the found synthesized expressions. The advantage is that equations for each output node can be mixed. Therefore, the best shapes for each output node can be selected and used together for correct behaviour. The training error for the example is equal to two misclassified items (it means 2.66 % error, 97.34 % success from all 75 training patterns) and testing error equal to five misclassified items (8 % error, 92 % success). The other cases were of similar errors.

$$y1 = -692.512\,x3\,\exp\left(\frac{-\,x3\left(\left(x1^{438.154}-257.159\right)\left(x3 - 2.05754\right)-832.086\right)\left(\exp(\exp(x2)) + \exp(x2)\right)^{-x4}}{\exp(x2) - \exp(-0.0010047\,x1)}\right) \tag{9.20}$$

$$y2 = -5.73269\times 10^{-48} - \exp\left(-\frac{0.0377017\,x1\left(-387.792^{320.911x4} + x4\right)}{x4}\right) + \frac{-713.828 + 700.061^{\frac{5.45627\times 10^{221}}{-953.763 + \exp(x2)}}}{x1} \tag{9.21}$$

$$y3 = \exp\left(-\frac{950.429}{\exp(x1 - 1.x3\,x4) + \frac{\exp\left(\exp\left(-\exp\left(-238.261 + x1\right)\right)^{\exp(\exp(57.0214 - x4))}\right)^{5.054576515780\times 10^{-391} + x2}}{-16.4604 + x3}}\right) \tag{9.22}$$

The following output (9.23) is from the training set.

$$\begin{aligned}\text{training output} = \{&\{1.,0.,0.\},\{1.,0.,0.\},\{1.,0.,0.\},\{1.,0.,0.\},\{1.,0.,0.\},\{1.,0.,0.\},\{1.,0.,0.\},\{1.,0.,0.\},\\ &\{1.,0.,0.\},\{1.,0.,0.\},\{1.,0.,0.\},\{1.,0.,0.\},\{1.,0.,0.\},\{1.,0.,0.\},\{1.,0.,0.\},\{1.,0.,0.\},\\ &\{1.,0.,0.\},\{1.,0.,0.\},\{1.,0.,0.\},\{1.,0.,0.\},\{1.,0.,0.\},\{1.,0.,0.\},\{1.,0.,0.\},\{1.,0.,0.\},\{1.,0.,0.\},\\ &\{0.,1.,0.\},\{0.,1.,0.\},\{0.,1.,0.\},\{0.,1.,0.\},\{0.,1.,0.\},\{0.,1.,0.\},\{0.,1.,0.\},\{0.,1.,0.\},\{0.,1.,0.\},\{0.,1.,0.\},\\ &\{0.,1.,0.\},\{0.,1.,0.\},\{0.,1.,0.\},\{0.,1.,0.\},\{0.,1.,0.\},\{0.,1.,0.\},\{0.,1.,0.\},\{0.,1.,0.\},\{0.,1.,0.\},\{0.,1.,0.\},\\ &\{0.,0.,1.\},\{0.,1.,0.\},\{0.,1.,0.\},\{0.,1.,0.\},\{0.,1.,0.\},\{0.,0.,1.\},\{0.,0.,1.\},\{0.,0.,1.\},\{0.,0.,1.\},\{0.,0.,1.\},\\ &\{0.,0.,1.\},\{0.,0.,1.\},\{0.,0.,1.\},\{0.,0.,1.\},\{0.,0.,1.\},\{0.,0.,1.\},\{0.,0.,1.\},\{0.,0.,1.\},\{0.,0.,1.\},\{0.,0.,1.\},\\ &\{0.,0.,1.\},\{0.,0.,1.\},\{0.,0.,1.\},\{0.,0.,1.\},\{0.,1.,1.\},\{0.,0.,1.\},\{0.,0.,1.\},\{0.,0.,1.\},\{0.,0.,1.\},\{0.,0.,1.\}\}\end{aligned} \tag{9.23}$$

It is visible that twenty-first item in the second group (iris versicolor) is misclassified as iris setosa and also twentieth item in the third group (iris setosa) is misclassified as partly iris setosa and partly iris versicolor, similar as in the one output and continues classification.

In the testing phase, we have obtained more misclassified items as can be seen in (9.24). Here, we have obtained one misclassified item in iris virginica (19th item), two in iris versicolor (3rd and 9th items) and also two in iris setosa (5th and 11th items).

$$
\begin{aligned}
\text{testing output} = \{&\{1.,0.,0.\},\{1.,0.,0.\},\{1.,0.,0.\},\{1.,0.,0.\},\{1.,0.,0.\},\{1.,0.,0.\},\{1.,0.,0.\},\{1.,0.,0.\},\\
&\{1.,0.,0.\},\{1.,0.,0.\},\{1.,0.,0.\},\{1.,0.,0.\},\{1.,0.,0.\},\{1.,0.,0.\},\{1.,0.,0.\},\{1.,0.,0.\},\{1.,0.,0.\},\{1.,0.,0.\},\{1.,1.,0.\},\\
&\{1.,0.,0.\},\{1.,0.,0.\},\{1.,0.,0.\},\{1.,0.,0.\},\{1.,0.,0.\},\{1.,0.,0.\},\{0.,1.,0.\},\{0.,1.,0.\},\{0.,0.,1.\},\{0.,1.,0.\},\\
&\{0.,1.,0.\},\{0.,1.,0.\},\{0.,1.,0.\},\{0.,1.,0.\},\{0.,1.,1.\},\{0.,1.,0.\},\{0.,1.,0.\},\{0.,1.,0.\},\{0.,1.,0.\},\{0.,1.,0.\},\\
&\{0.,1.,0.\},\{0.,1.,0.\},\{0.,1.,0.\},\{0.,1.,0.\},\{0.,1.,0.\},\{0.,1.,0.\},\{0.,1.,0.\},\{0.,1.,0.\},\{0.,1.,0.\},\{0.,1.,0.\},\\
&\{0.,1.,0.\},\{0.,0.,1.\},\{0.,0.,1.\},\{0.,0.,1.\},\{0.,0.,1.\},\{0.,1.,1.\},\{0.,0.,1.\},\{0.,0.,1.\},\{0.,0.,1.\},\{0.,1.,0.\},\\
&\{0.,1.,1.\},\{0.,0.,1.\},\{0.,0.,1.\},\{0.,0.,1.\},\{0.,0.,1.\},\{0.,0.,1.\},\{0.,0.,1.\},\{0.,0.,1.\},\{0.,0.,1.\},\\
&\{0.,0.,1.\},\{0.,0.,1.\},\{0.,0.,1.\},\{0.,0.,1.\},\{0.,0.,1.\},\{0.,0.,1.\},\{0.,0.,1.\}\}
\end{aligned}
\tag{9.24}
$$

The obtained number of training and testing errors are comparable with errors obtained within other approaches as artificial neural networks and others for both cases.

9.4 Conclusion

The chapter dealt with the usage of analytic programming, one of the evolutionary symbolic regression technique for two applications—control law synthesis for stabilizing of deterministic chaotic system and classification with pseudo neural networks.

The result from the performed simulations reinforces the argument that AP is able to solve these difficult problems.

AP produced new synthesized control laws in a symbolic way securing desired behaviour of a selected chaotic system—Hénon map. Precise and fast stabilization gives weight to the argument that AP is a powerful symbolic regression tool which is able to strictly and precisely follow the rules given by the cost function and synthesize some kind of a feedback controller for a chaotic system.

The same is valid for the case of pseudo neural networks. Within this approach, classical optimization of the structure or weights was not performed. The proposed technique is based on symbolic regression with evolutionary computation. It synthesizes a whole structure in symbolic form without a prior knowledge of the ANN structure or transfer functions. It means that the relation between inputs and output(s) is synthesized. As can be seen from the result section, such approach is promising.

The research never ends. The question of energy costs and more precise stabilization will be included into the future research together with consideration of observed critical points, the development of better cost functions, a different AP data set and performing of numerous simulations to obtain more results and produce better statistics, thus to confirm the robustness of these approaches to control law and pseudo neural nets synthesizes.

Acknowledgement This work was supported by the Ministry of Education, Youth and Sports of the Czech Republic within the National Sustainability Programme project No. LO1303 (MSMT-7778/2014) and also by the European Regional Development Fund under the project CEBIA-Tech No. CZ.1.05/2.1.00/03.0089, further it was supported by Grant Agency of the Czech Republic—GACR 588P103/15/06700S.

References

1. Back T, Fogel DB, Michalewicz Z (1997) Handbook of evolutionary algorithms. Oxford University Press. ISBN: 0750303921
2. Deugo D, Ferguson D (2004) Evolution to the Xtreme: evolving evolutionary strategies using a meta-level approach. In: Proceedings of the 2004 IEEE congress on evolutionary computation. IEEE Press, Portland, Oregon, pp 31–38
3. Dioşan L, Oltean M (2009) Evolutionary design of evolutionary algorithms. Genet Program Evolvable Mach 10(3):263–306
4. Edmonds B (2001) Meta-genetic programming: co-evolving the operators of variation. Elektrik 9(1):13–29
5. Eiben AE, Michalewicz Z, Schoenauer M, Smith JE (2007) Parameter control in evolutionary algorithms. Springer, pp 19–46
6. Fausett LV (1993) Fundamentals of neural networks: architectures, algorithms and applications. Prentice Hall, ISBN: 9780133341867
7. Fekiac J, Zelinka I, Burguillo JC (2011) A review of methods for encoding neural network topologies in evolutionary computation. In: ECMS 2011, Krakow, Poland, ISBN: 978-0-9564944-3-6
8. Fisher RA (1936) The use of multiple measurements in taxonomic problems. Ann. Eugenics 7 (2):179–188. doi:10.1111/j.1469-1809.1936.tb02137.x
9. Gurney K (1997) An introduction to neural networks. CRC Press, ISBN: 1857285034
10. Hertz J, Kogh A, Palmer RG (1991) Introduction to the theory of neural computation. Addison-Wesley
11. Hilborn RC (2000) Chaos and nonlinear dynamics: an introduction for scientists and engineers. Oxford University Press, ISBN: 0-19-850723-2
12. Jones DF, Mirrazavi SK, Tamiz M (2002) Multi-objective meta-heuristics: an overview of the current state-of-the-art. Eur J Oper Res 137(1):1–9, ISSN: 0377-2217
13. Just W (1999) Principles of time delayed feedback control. In: Schuster HG (ed) Handbook of chaos control. Wiley-Vch, ISBN: 3-527-29436-8
14. Kalczynski PJ, Kamburowski J (2007) On the NEH heuristic for minimizing the makespan in permutation flow shops. Omega 35(1):53–60
15. Kordík P, Koutník J, Drchal J, Kovářík O, Čepek M, Šnorek M (2010) Meta-learning approach to neural network optimization. Neural Netw 23(4):568–582, ISSN: 0893-6080
16. Koza JR et al (1999) Genetic programming III; darwinian invention and problem solving. Morgan Kaufmann Publisher, ISBN: 1-55860-543-6
17. Koza JR (1998) Genetic programming. MIT Press, ISBN: 0-262-11189-6
18. Kwon OJ (1999) Targeting and stabilizing chaotic trajectories in the standard map. Phys Lett A 258:229–236
19. Lampinen J, Zelinka I (1999) New ideas in optimization—mechanical engineering design optimization by differential evolution, vol 1. McGraw-hill, London, 20p, ISBN: 007-709506-5
20. Machine learning repository with Iris data set http://archive.ics.uci.edu/ml/datasets/Iris
21. Murty KG (1983) Linear programming. Wiley, New York, ISBN: 0-471-09725-X
22. Murty KG (1988) Linear complementarity, linear and nonlinear programming, Sigma series in applied mathematics. Heldermann Verlag, Berlin, ISBN: 3-88538-403-5
23. O'Neill M, Ryan C (2003) Grammatical evolution. Evolutionary automatic programming in an arbitrary language. Kluwer Academic Publishers, ISBN: 1402074441
24. Oplatkova Z (2009) Metaevolution: synthesis of optimization algorithms by means of symbolic regression and evolutionary algorithms. Lambert Academic Publishing Saarbrücken, ISBN: 978-3-8383-1808-0
25. Oplatková Z, Zelinka I (2009) Investigation on evolutionary synthesis of movement commands, modelling and simulation in engineering, vol 2009, Article ID 845080, 12p. Hindawi Publishing Corporation, ISSN: 1687-559

26. Oplatkova Z, Senkerik R, Zelinka I, Holoska J (2010) Synthesis of control law for Chaotic Henon system—preliminary study, ECMS 2010, Kuala Lumpur, Malaysia, pp 277–282, ISBN: 978-0-9564944-0-5
27. Oplatkova Z, Senkerik R, Belaskova S, Zelinka I (2010) Synthesis of control rule for synthesized chaotic system by means of evolutionary techniques, Mendel 2010, Brno, Czech Republic, pp 91–98, ISBN: 978-80-214-4120-0
28. Ott E, Greboki C, Yorke JA (1990) Controlling chaos. Phys Rev Lett 64:1196–1199
29. Price K, Storn RM, Lampinen JA (2005) Differential evolution: a practical approach to global optimization, (Natural computing series), 1st edn. Springer
30. Price K, Storn R (2001) Differential evolution homepage. http://www.icsi.berkeley.edu/~storn/code.html, [Accessed 29/02/2012]
31. Pyragas K (1992) Continuous control of chaos by self-controlling feedback. Phys Lett A 170:421–428
32. Pyragas K (1995) Control of chaos via extended delay feedback. Phys Lett A 206:323–330
33. Senkerik R, Zelinka I, Navratil E (2006) Optimization of feedback control of chaos by evolutionary algorithms. In: Proceedings 1st IFAC conference on analysis and control of chaotic systems, Reims, France, pp 97–102
34. Senkerik R, Zelinka I, Davendra D, Oplatkova Z (2009) Utilization of SOMA and differential evolution for robust stabilization of chaotic logistic equation. Comput Math Appl 60(4):1026–1037
35. Senkerik R, Oplatkova Z, Zelinka I, Davendra D, Jasek R (2010) Synthesis of feedback controller for chaotic systems by means of evolutionary techniques. In: Proceeding of fourth global conference on power control and optimization, Sarawak, Borneo (2010)
36. Smith J, Fogarty T (1997) Operator and parameter adaptation in genetic algorithms. Soft Comput 1(2):81–87
37. Voutsinas TG, Pappis CP (2010) A branch and bound algorithm for single machine scheduling with deteriorating values of jobs. Math Comput Model 52(1–2):55–61
38. Wasserman PD (1980) Neural computing: theory and practice. Coriolis Group, ISBN: 0442207433
39. Zelinka et al (2004) Analytical programming—a novel approach for evolutionary synthesis of symbolic structures, in Kita E.: evolutionary algorithms, InTech 2011, ISBN: 978-953-307-171-8
40. Varacha P, Zelinka I, Oplatkova Z (2006) Evolutionary synthesis of neural network, Mendel 2006—12th international conference on softcomputing, Brno, Czech Republic, 31 May–2 June 2006, pp 25–31, ISBN: 80-214-3195-4
41. Zelinka I,Oplatkova Z, Nolle L (2005) Boolean symmetry function synthesis by means of arbitrary evolutionary algorithms-comparative study. Int J Simul Syst Sci Technol 6(9):44–56, ISSN: 1473-8031
42. Zelinka I, Senkerik R, Navratil E (2009) Investigation on evolutionary optimization of chaos control. Chaos Solitons Fractals 40(1):111–129
43. Zelinka I, Guanrong Ch, Celikovsky S (2008) Chaos synthesis by means of evolutionary algorithms. Int J Bifurcat Chaos 18(4):911–942

Chapter 10
On Practical Automated Engineering Design

Lars Nolle, Ralph Krause and Richard J. Cant

Abstract In engineering, it is usually necessary to design systems as cheap as possible whilst ensuring that certain constraints are satisfied. Computational optimization methods can help to find optimal designs automatically. However, it is demonstrated in this work that an optimal design is often not robust against variations caused by the manufacturing process, which would result in unsatisfactory product quality. In order to avoid this, a meta-method is used in here, which can guide arbitrary optimization algorithms towards more robust solutions. This was demonstrated on a standard benchmark problem, the pressure vessel design problem, for which a robust design was found using the proposed method together with self-adaptive stepsize search, an optimization algorithm with only one control parameter to tune. The drop-out rate of a simulated manufacturing process was reduced by 30 % whilst maintaining near-minimal production costs, demonstrating the potential of the proposed method.

Keywords Automated engineering design · Robust solution · Optimization · Parameter tuning · Self-adaptive stepsize search · Pressure vessel problem

---

L. Nolle (✉)
Department of Engineering Science, WE Applied Computer Science,
Jade University of Applied Science, Friedrich-Paffrath-Straße 101,
26389 Wilhelmshaven, Germany
e-mail: lars.nolle@jade-hs.de

R. Krause
Siemens AG, Energy Management, EM MS S SO PR, Mozartstrasse 31c,
91052 Erlangen, Germany
e-mail: ralph.krause@siemens.com

R.J. Cant
School of Science and Technology, Nottingham Trent University,
Clifton Lane, NG11 8NS Nottingham, UK
e-mail: richard.cant@ntu.ac.uk

K. Al-Begain and A. Bargiela (eds.), *Seminal Contributions to Modelling and Simulation*, Simulation Foundations, Methods and Applications,
DOI 10.1007/978-3-319-33786-9_10

10.1 Engineering Design

The Oxford English Dictionary defines engineering as the "… branch of science and technology concerned with the development and modification of engines […], machines, structures, or other complicated systems and processes using specialized knowledge or skills…" [40]. In other words, engineering uses sound scientific methods and technologies to solve real-world problems and develop new products and services.

A typical task for an engineer is to design a system that is capable of fulfilling a desired function. This could be, for example, a bridge that spans over a defined distance that can resist a predefined maximum load. There is not one 'right' design that solves the problem. Instead many different designs are feasible. However, some designs are preferable to others, for example because they are cheaper to build or easier to manufacture. The challenge for the engineer is to choose the 'best' design from all of the feasible designs, where 'best' is defined in relation to the specific case by interested parties, i.e. stakeholders.

Figure 10.1 shows a simple yet typical example of an engineering design problem: point *B* has to withstand a force *F* at distance l_1 from point *A* without exceeding a maximum displacement ε. This can be achieved by putting a beam, also called a *member*, between points *A* and *B* (Fig. 10.1a). Such a design, however, could be prone to plastic deformation under load if the cross-sectional area of the beam is not sufficiently large enough. On the other hand, if the cross-sectional area is unnecessarily large, the beam uses a lot of material and hence is very expensive and heavy. In order to save material and thus make the structure lighter, usually the load is distributed by adding additional supporting beams (Fig. 10.1b). In structural engineering, it is common to treat such structures as trusses, which consist of two-force members only [10]. This means that forces apply only to the endpoints of a member and the joints between the members are treated as revolutes, i.e. they do

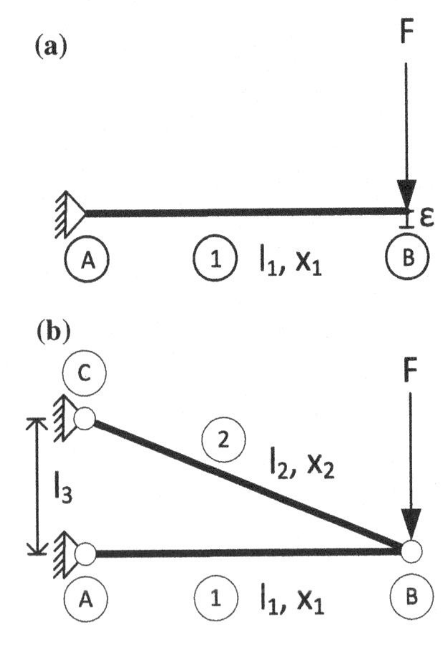

Fig. 10.1 Required function and practical implementation. **a** Functional requirements, **b** practical implementation

not carry moments (torques). Therefore, at least one more member, member 2 in Fig. 10.1b, is necessary to hold point B in position.

The length l_1 of member 1 is defined by the problem and therefore fixed but length l_2 of member 2 can be chosen freely by the designer within limits. This length then determines the position of point C and therewith the length l_3 (Fig. 10.1b). The material used has a specific Young's module E and a density ρ. The volume of a member, which is given by its length l multiplied by its cross-section x, and its density ρ determines the weight of the member. Each design can be fully described by its design vector $\vec{d}$. Equation 10.1 shows the design vector for the example above:

$$\vec{d} = \begin{Bmatrix} l_1 \\ l_2 \\ x_1 \\ x_2 \\ E \\ \rho \end{Bmatrix} \qquad (10.1)$$

Some of the design variables are predetermined by the problem (l_1, E, ρ in the example) and hence are called the *preassigned parameters* [44]. They cannot be changed. However, others can be chosen freely by the designer (l_2, x_1, x_2 in the example) and these are called *design* or *decision variables*.

A system described by its design vector can be judged in terms of goodness using a so-called *objective function* $f(\vec{d})$. In structural engineering, this objective function usually relates to the costs of a particular design, which is reflected in the weight of the resulting structure [51]. For example, the weight of the truss in Fig. 10.1 can be calculated as follows:

$$f(\vec{d}) = l_1 x_1 \rho + l_2 x_2 \rho \qquad (10.2)$$

If better designs are associated with smaller function values, the objective function is usually referred to as the *cost function*. If, on the contrary, better designs are associated with larger objective function values, the objective function is usually referred to as the *fitness function*.

10.1.1 Optimal Design

As mentioned before, the challenge for the engineer is to find the optimal design. The objective function can be used to compare candidate designs with the aim of selecting the best one, i.e. the lightest one in the example. The optimal design is the design $\vec{d}'$ that minimizes the objective function $f(\vec{d})$:

$$\vec{d}' = \underset{\vec{d}' \in C}{\arg\min}\ f(\vec{d}) \qquad (10.3)$$

In Eq. 10.3, C denotes the space of all feasible designs. If the objective function is differentiable twice and the design space C is unconstrained, calculus-based methods [6] can be applied to calculate the optimal solution. However, often, this is not the case, for example because of the complexity of the system and hence the complexity of the objective function. In such a case, a common approach to design optimization is stepwise refinement: based on an initial feasible design, a human expert, i.e. the engineer, makes small adjustments to the design and evaluates its fitness. If the fitness increases the adjustments are kept; otherwise the adjustments are discarded. This is then carried out iteratively until a design is arrived at that satisfies the requirements (Fig. 10.2).

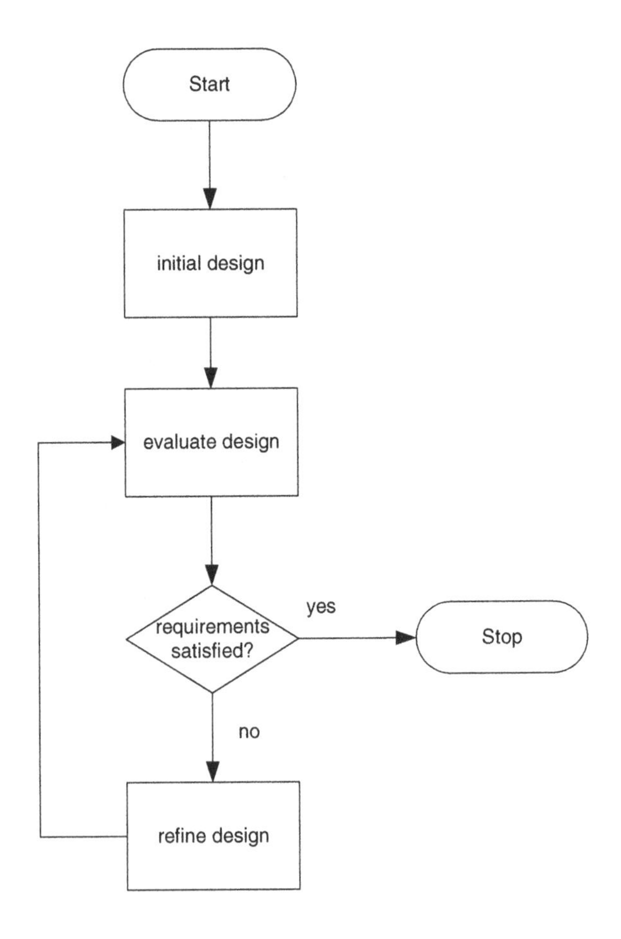

Fig. 10.2 The design process

10.1.2 Constraint Optimization

However, not every design is actually feasible; the objective function in Eq. 10.2 would return the lowest objective value if both x_1 and x_2 were equal to or smaller than zero. Obviously, such a design is impossible. Therefore, the cross-sections have to be chosen to be as small as possible. But if the cross-sections are too small, the members cannot withstand external load and the structure will fail. Hence, the design space, i.e. the collection of all possible designs, is limited by constraints. Thus, the optimization problem can be re-formulated as follows [44]:

Find $\vec{d} = \begin{Bmatrix} x_1 \\ x_2 \\ x_3 \\ \vdots \\ x_n \end{Bmatrix}$ that minimizes or maximizes $f(\vec{d})$ subject to the following constraints:

$$\begin{aligned} g_i(\vec{d}) &\leq 0, \quad i = 1, 2, \ldots, m \\ h_i(\vec{d}) &= 0, \quad i = 1, 2, \ldots, p \end{aligned} \tag{10.4}$$

In Eq. 10.4, $g_i(\vec{d})$ denotes *inequality* constraints and $h_i(\vec{d})$ denotes *equality* constraints. The number of decision variables n of a design and the number of inequality or equality constraints (m respectively p) are usually not related to each other for a given problem. If one or more inequality constraints or equality constraints are violated the design is not feasible. In order to accommodate constraints in the optimization process, equalities are usually handled by converting them into inequality constraints [11]:

$$h_{m+i} \equiv \delta - \left| h_i(\vec{d}) \right| \geq 0 \tag{10.5}$$

In Eq. 10.5, δ is a small positive value. This conversion increases the total number of inequalities to $q = m + p$, and allows for handling inequalities and equalities using the same process. For constraint optimization problems, a penalty term $p(\vec{d})$ is added to the objective function $f(\vec{d})$:

$$F(\vec{d}) = f(\vec{d}) + p(\vec{d}) \tag{10.6}$$

Instead of using the objective function value, the value returned by $F(\vec{d})$ is used in the optimization process. In Eq. 10.6, the penalty term is computed as follows:

$$p(\vec{d}) = \sum_{i=1}^{m} R_i(g_i(\vec{d})) \tag{10.7}$$

Here, $R_i(g_i(\vec{d}))$ is a penalty function for the ith constraint, i.e. a function that returns a non-zero value if the ith constraint is violated or otherwise zero. This has the effect that a design that violates a constraint is penalized by its fitness decreasing in the case of maximizing or its costs increasing in the case of minimizing.

Choosing a suitable penalty function for a particular optimization problem is not a straightforward task. In the simplest case, the penalty function adds a constant number to the objective function for each constraint violated. However, many researchers have developed more sophisticated methods for calculating penalty values. For example, [11] proposed a constraint handling method that does not require any fixed penalty parameter. [47] developed stochastic ranking for constraint handling. [28] introduced the minimum penalty coefficient method. A comprehensive survey of the most popular constraint handling techniques can be found in [7].

By using a penalty function, a constraint optimization problem is transformed into an unconstraint problem, and hence can be solved automatically as explained in the next section.

10.2 Automated Design

The drawback of the design process outlined above is that it relies on human expertise and experience. For example, if the engineer is inexperienced, they might limit themselves subconsciously to a certain area of the design space due to their personal biases. Therefore, an automated approach to the optimization problem that removes this bias would be of benefit.

Figure 10.3a shows the manual approach to design optimizations whereas Fig. 10.3b presents a computerized automated optimization approach; a computational optimization algorithm is used instead of a human engineer in a closed *optimization loop* to experiment iteratively on the system in order to find the best solution [37].

Starting with an initial design x_0, the fitness of that design is evaluated using an objective function. Based on the fitness, an optimization algorithm makes adjustments to the decision variables, resulting in a new design vector x_1. The new design is then evaluated and the new fitness respectively cost information is used by the algorithm to further adjust the design vector. Depending on the algorithm used and the number of iterations, the objective value will improve until it converges towards an optimum value.

Automated design optimization appeared in the early 1970s when computers became more readily available to researchers and engineers. For examples see [54] or [41]. Since then, researchers have developed a large variety of optimization algorithms, such as simulated annealing [25], genetic algorithms [18] and particle swarm optimization [24], to name a few, and applied them to engineering design problems. Recent examples include the design of rotor-bearing systems [30] and the

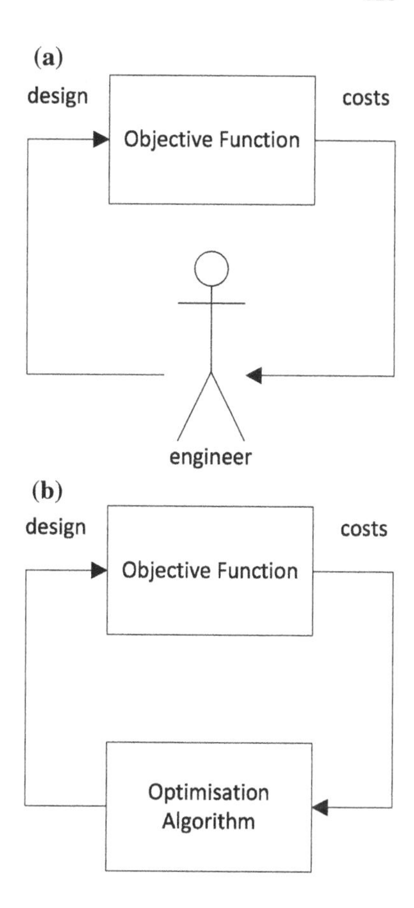

Fig. 10.3 Optimization loop for optimal design. **a** Manual design process, **b** automated design process

design of sub-structures for offshore wind turbines [57]. All of these applications have shown that automated design optimization is capable of solving complex design problems. However, these techniques have not yet been widely adopted by the engineering industry [37]. The reasons for this will be outlined later.

10.2.1 Current Research

Many computational optimization algorithms, or direct search methods, have been proposed in the past. Early algorithms included, for example hill climbing [4], simulated annealing [25, 33], the simplex method [35], evolutionary strategies [45] and genetic algorithms [16, 18]. This was followed by particle swarm optimization [24], ant colony optimization [14], differential evolution [53] and self-adaptive stepsize search [39], amongst others. Later, it became popular to combine these methods with local search schemes to improve the search capabilities of the algorithms, especially with genetic algorithms.

Many new schemes have since been proposed. For example, [36] introduced a new hybrid algorithm, called hybrid particle swarm branch-and-bound (HPB), which uses particle swarm optimization (PSO) as the global optimization technique and combines it with branch-and-bound (BB) to solve mixed discrete nonlinear programming problems (MDNLP) [26]. PSO showed good global but slow local search abilities and was therefore combined with the BB technique, which features fast local search ability. The combination of these two search techniques exhibits improved optimization accuracy and reduced demand for computational resources [56]. The HPBs hybridization phase uses mainly a selective temporary varying from PSO to BB when the current optimum could be improved [13]. HPB can handle random and discrete constraints without the need for parameterizing a penalty function [22]. This architecture makes the HPB simple, generic and easy to implement. Nema et al. were also able to show that the combination of both search techniques results in better solutions and a lower number of function evaluations than using BB or PSO separately.

Most recently, for example, [27] proposed an improved, chemical reaction optimization (CRO) method, called OCRO, to solve global numerical optimization problems. The CRO method mimics the interactions between the molecules in chemical reactions to find a low energy stable state. The proposed algorithm is a hybrid of two local search operators in CRO and a quantization orthogonal crossover (QOX) search operator. An advantage of this approach is that no complicated operators are necessary, which makes it easy to implement in various real-world problems. The experimental results have shown that the algorithm is effective and fast in solving common optimization problems but is less efficient in solving low-dimensional functions.

Wang et al. [55] have tried a different approach and introduced a hybrid algorithm combining genetic particle swarm optimization with advanced local search ability (L-GPS) and a quasi-newton genetic algorithms (GA) with limited memory. Because of its wide range search ability, GA is used to search global minima/maxima. The global extremes are then further investigated in terms of local extremes with an algorithm that shows good local search abilities (L-GPS). The hybrid algorithm was then evaluated using numerical experiments that showed that the hybrid algorithm is much faster and more efficient than most stochastic methods.

Another recent approach has been published by Zhou et al. [58], who proposed in their paper a new hybrid glowworm swarm optimization algorithm, named the hybrid glowworm swarm optimization algorithm (HGSO), to solve constrained optimization problems. A fish school algorithm (FSA) in which predatory behaviour is exhibited is integrated with a glowworm swarm optimization to obtain the improved glowworm swarm optimization (IGSO). Here, the IGSO algorithm is integrated with the differential evolution method (DE) to establish the HGSO. The local search strategy used to escape from the local minimum is based on simulated annealing (SA) and is applied to the best solution found by the IGSO. The experimental results have shown that the HGSO is very effective in terms of efficiency, reliability and precision.

In their recent paper, [31] suggested a new hybrid method based on particle swarm optimization (PSO), artificial ant colony optimization (ACO) and 3-opt algorithm to solve the travelling salesman problem (TSP). In this approach the PSO is used to determine parameters that affect the performance of the ACO. The 3-opt algorithm is subsequently used to remove the local solution. The performance of this method was investigated by altering the average route length or the effective number of ants in the ACO. The experimental results show that the fewer the number of ants, the better the performance of the proposed method.

Jayaprakasam et al. [21] suggested a new algorithm, PSOGSA-E, which combines the local search ability of the gravitational search algorithm (GSA) with the social thinking of PSO. The new hybrid algorithm improves the beam pattern of collaborative beamforming to achieve lower side lobes by optimizing the weight vector of the array elements. By using exploration and exploitation, a global optimum is determined by the hybrid algorithm and the problem of premature convergence is alleviated compared to the legacy optimization methods.

Jamshidi et al. [20] introduced an optimization method based on a set of performance indices to determine the effectiveness and robustness of a supply chain. These indices are categorized into qualitative and quantitative indices to optimize supply chain parameters like costs or waiting time. An artificial immune system (AIS) was utilized as a meta-heuristic method and it was then combined with the Taguchi method to optimize the problem. It was shown that this novel hybrid method leads to more accurate minimum average costs and less standard deviation. In computational tests, it was also shown that the procedure was effective and efficient and could be used for a variety of optimization problems.

Another novel hybrid algorithm of teaching–learning-based optimization (TLBO) and a DE was introduced by Pholdee et al. [42]. The algorithm was used to optimize a strip coiling process and it has been shown that it can also handle large-scale design problems for industrial applications. The optimal processing parameters are determined by minimization of the proposed objective function with a limited number of function evaluations. The hybrid algorithm was tested and compared with ten already existing EA strategies and it was able to outperform the other EAs in terms of efficiency and computational time.

In their recent paper, [32] described the application of bio-inspired methods to designing logic controllers using genetic algorithms (GA), particle swarm methods (PSO) and hybrid PSO-GA methods. Some of these optimization methods are based on populations of solutions and produce a set of solutions for each iteration step. All of these different solutions are selected, combined and replaced to find the optimized solution. This process requires much more computing time than other meta-heuristic methods and a greater numbers of iterations. Martínez-Soto's aim was to develop more aggressive methods by combining population-based algorithms with single solution algorithms. The experimental procedure was realized for each method individually (GA, PSO) and for both methods combined (PSO-GA). Afterwards the results obtained using each method individually were compared with the results obtained using both methods combined.

Arnaud and Poirion [1] proposed a combinatorial optimization algorithm matched to a continuous optimization problem that is related to uncertain aero elastic optimization. The algorithm optimizes the sum of the expected value by stochastic annealing of the objective function. Additionally a penalized term regards the confidence interval bandwidth of the estimator. An advantage of this algorithm is that it chooses the number of samples used for estimator construction. By not using a fixed number of samples, the computational efficiency is improved compared to a classical annealing algorithm. The test problem shows that the algorithm can be used efficiently for low dimension optimization problems under uncertainty with no objective function gradient. Compared to gradient-based approaches, the computation time of the proposed algorithm is more important so that it can be inefficient for high dimension industrial application.

Another interesting strategy was introduced by Congedo et al. [9], who proposed a multi-scale strategy called the simplex 2 method (S2M), to minimize the cost of robust optimization procedures. It is based on simplex tessellation of uncertainty as well as of the design space. The algorithm coupled the simplex stochastic collocation (SSC) method employed for uncertainty quantification with the Nelder–Mead optimization algorithm [35]. The robustness of the S2M method is caused by using a coupled stopping criterion and a high-degree polynomial interpolation with P-refinement on the design space. This method has been applied to various algebraic benchmark test cases like the Rosenbrock problem, for which it was able to reduce the global number of deterministic evaluations by about 50 %. Another test case was the short column problem, which was treated using the S2M method and a classical technique called polynomial chaos expansion (PCE) with genetic algorithms (GA). The comparison of both methods has shown that S2M is able to gain one order of magnitude in terms of convergence order while maintaining the same level of error.

All of the latest methods presented here are hybrid algorithms. That means that they combine two or more optimization algorithms to improve either effectiveness or efficiency. In the former case, they aim at increasing the quality of the solutions, i.e. finding solutions with a higher fitness value in the case of maximization or lower cost values in the case of minimization. In the latter case they aim at minimizing the computational costs, which are caused by objective function evaluation.

10.2.2 Practical Problems

The tuning of optimization algorithms, and hybrid algorithms in particular, presents problems to practitioners who want to use direct search techniques as black-box methods for practical engineering applications: All of these algorithms have a number of control parameters that need to be carefully adjusted to the optimization problem at hand. This constitutes a meta-optimization problem, i.e. the optimum set of control parameters has to be found, usually empirically, before the main optimization process can start. This fine-tuning of the control parameters requires a lot of experience with optimization algorithms and usually a large number of experiments.

For example, the basic GA has seven parameters that need to be tuned before the actual optimization can be carried out. Due to combinatorial explosion and the continuous nature of some of the parameters, not every combination can be tried in an exhaustive search. Using unsuitable settings can easily lead to sub-optimal solutions.

Engineers are often not very experienced with search algorithms and hence often struggle to find suitable settings within the limited time available for tuning. Ideally, engineers would like to use an algorithm that has no control parameter at all. This has led to the development of self-adaptive stepsize search (SASS), an algorithm that has only one control parameter [2, 3, 37, 39]. It has been demonstrated in practical applications that SASS can be used without the need for tuning control parameters [12, 38, 50].

Although SASS does not require parameter tuning, apart from selecting the number of particles in the population, the algorithm might take a longer time to reach the global optimum and hence is not as efficient as the latest hybrid algorithms outlined above. However, for practical applications, this is compensated for by the fact that no time needs to be spent experimenting with parameter settings.

10.3 Case Study

In engineering, designing a system that satisfies all of the user's requirements, without violating problem specific constraints, usually requires determining the system's design variable values in such a way that the resulting design is optimal in turns of cost function. Usually, the design space is too vast to evaluate every possible design and hence only a subset of the design space can be considered. An interesting standard benchmark problem from the field of mechanical engineering, which was introduced by [49], is the pressure vessel problem. It has been used as a standard problem by many researchers. Current examples include [17, 23, 29, 48]. Although the problem only consists of four decision variables, it is not a trivial problem, as it involves the optimization of both discrete and continuous design variables under a relatively large number of constraints. The problem is used as a test-bed here and is therefore explained in more detail in the next section.

10.3.1 Pressure Vessel Problem

The pressure vessel consists of a simple cylinder (shell) with each end capped by a hemispherical shell (head), as illustrated in Fig. 10.4. There are four design variables, which can be chosen by the engineer: the thickness $x1$ of the shell, the thickness $x2$ of the heads, the inner radius $x3$ and the length $x4$ of the shell.

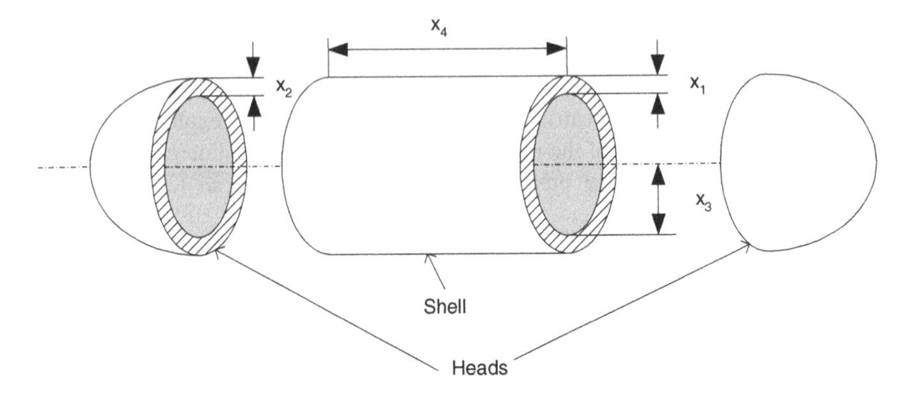

Fig. 10.4 Pressure vessel design problem

Table 10.1 Design parameter types and ranges for the pressure vessel problem

Design variable	Definition	Variable type	Thickness increments
x1	Thickness of cylinder	Discrete	0.0625 in.
x2	Thickness of sphere	Discrete	0.0625 in.
x3	Inner radius of cylinder	Continuous	N/A
x4	Length of cylinder	Continuous	N/A

The variables x1 and x2 are both discrete and vary in multiples of 0.0625 inches (please note: Sandgren used Imperial units), and variables x3 and x4 are both continuous (see Table 10.1).

The design has to satisfy four constraint functions and an allowed range for each design variable. The design and constraint functions and the ranges of the design variables are given in the following equations and inequalities:

$$\begin{aligned} g_1(\mathbf{x}) &= 0.0193x_3 - x_1 \leq 0 \\ g_2(\mathbf{x}) &= 0.00954x_3 - x_2 \leq 0 \\ g_3(\mathbf{x}) &= 1,296.\ 000 - \pi x_3^2 x_4 - 4/3\pi x_3^3 \leq 0 \\ g_4(\mathbf{x}) &= x_4 - 240 \leq 0 \\ 0.0625 &\leq x_1 \leq 6.1875 \\ 0.0625 &\leq x_2 \leq 6.1875 \\ 10 &\leq x_3 \leq 200 \\ 10 &\leq x_4 \leq 200 \end{aligned} \tag{10.8}$$

The aim is to minimize the total cost of the materials used and also the production costs, which consist of the costs of forming and welding the pressure vessel. Equation (10.9) provides the fitness function *f*(*x*), which was introduced by [49]. It comprises the four design variables x = (x1, x2, x3, x4).

Table 10.2 Comparison of results

	PSO	HPB	SASS
Fitness	6,059.7143	6,059.6545	6,059.7143
x1	0.8125	0.8125	0.8125
x2	0.4375	0.4375	0.4375
x3	42.09845	42.09893	42.09845
x4	176.6366	176.6305	176.5366

$$\begin{aligned} f(\mathbf{x}) =& 0.6224\, x_1x_3x_4 + 1.7781\, x_2x_3^2 \\ &+ 3.1661\, x_1^2x_4 + 19.84\, x_1^2x_3 \end{aligned} \tag{10.9}$$

Previous methods used are, amongst many others, Branch-and-Bound (BB) [49], Evolutionary Programming (EP) [5], Genetic Algorithm [8], Particle Swarm Optmization (PSO) [19], Hybrid Particle Swarm Branch-and-Bound (HPB) [36] and Self-Adaptive Stepsize Search (SASS) [39]. Table 10.2 shows the best designs, as found by the three best methods, and the associated fitness values as reported by [39]. As can be seen, the fitness (costs) could be reduced dramatically, compared to Sandgren's original value of 7982.5.

This example clearly showed that computational optimization methods are capable of finding very good solutions for engineering applications in terms of fitness function values. However, when implementing optimal designs physically, engineers are often confronted with a number of problems, which are outlined below.

10.3.2 Problems with Optimal Solutions

As can be seen from the pressure vessel example, computational optimization methods are capable of finding very good solutions for engineering applications in terms of fitness function values. However, solutions presented in the literature often use design parameter values that have many decimal places. This means, in reality, that those solutions cannot actually be manufactured because of the tolerances of both the materials involved and the manufacturing processes.

For the pressure vessel problem, for example, the solutions presented in the literature show up to five places after the decimal point. Since the manufacturing process cannot achieve the required accuracy, the parameter values are slightly different and hence result in a different fitness value. For example, using a modern CNC machine, tolerances of ± 0.005 mm are common.

Also, controlling the dimensions of the heads is especially difficult [43]. Other unavoidable causes of error are the thickness tolerances for plate rolled on a plate mill (Standards [52]). Table 10.3 shows the combined upper and lower tolerances for the pressure vessel design problem.

Table 10.3 Upper and lower tolerances for pressure vessel

Tolerance (in.)	x_1	x_2	x_3	x_4
Lower limit	−0.0118	−0.0118	−0.1969	−0.2756
Upper limit	0.0472	0.0472	0.1969	0.2756

Table 10.4 Fitness after applying upper and lower tolerances to the original design solutions

Method	Original fitness	Fitness for lower tolerance values	Fitness for upper tolerance values
PSO	6,059.7143	N/A (g1 and g3 violated)	6579.68
HPB	6,059.6545	N/A (g1 and g3 violated)	6579.60
SASS	6,059.7143	N/A (g1 and g3 violated)	6579.678

The upper and lower tolerance values were added to the design solutions presented in Table 10.2. Table 10.4 shows how the fitness values changed when the upper and lower tolerances were taken into account.

As can be seen from Table 10.4, the fitness decreases dramatically for all of the designs using the upper tolerance values, with an error of 8.6 %. Even worse, when using the lower tolerances, all of the designs are unfeasible because they violate constraints g_1 and g_3 and hence would not be fit for purpose!

There are two aspects of robust optimization. The first is related to the narrowness of the global optimum for unconstraint optimization. Figure 10.5 shows an example of a fitness landscape for a one-dimensional optimization problem. In the example, there are two maxima, one at $x = 5$ and the other at $x = 10$. The global optimum is located at $x = 5$. However, the peak is too narrow, and slight deviations in implementing the solution will cause a significant drop of the fitness value. The local optimum at $x = 10$ does not offer the same fitness, but slight deviations in x do not cause a dramatic drop in fitness when implementing the solution. If the fitness were acceptable, this would be the preferred solution for an engineering application. For a practical application, it would be more desirable to find a robust solution rather than the global optimum.

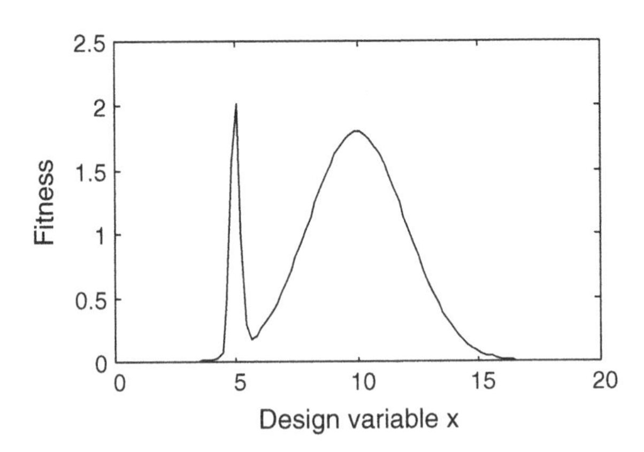

Fig. 10.5 Example of a fitness landscape for a one-dimensional optimization problem

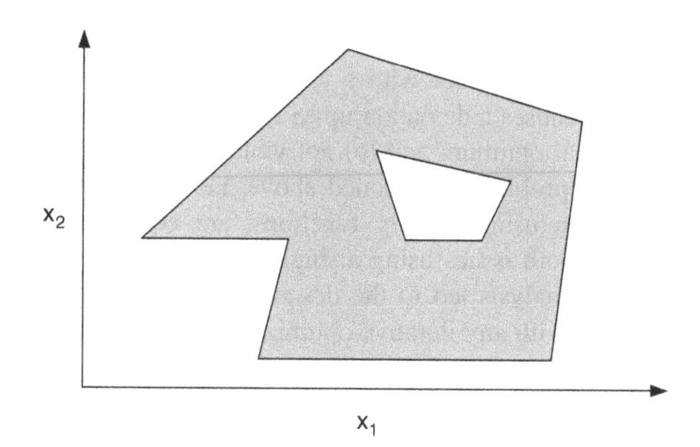

Fig. 10.6 Example of an input space for a two-dimensional constraint optimization problem

The other problem relates to constraint optimization. Here, constraints define areas in the input space which do not contain feasible solutions. Figure 10.6 shows an example of an input space for a two-dimensional constraint optimization problem.

In the example, there are two design variables, $x1$ and $x2$. However, only combinations of $x1$ and $x2$ that lie in the grey area are feasible, i.e. allowed, whereas combinations that lie in the white areas are not feasible. The white areas are defined by the constraints. The more constraints there are, the more complicated the shape of the allowed area(s) can be and hence the more difficult the optimization problem.

It is a well-known fact that optimum solutions are normally located at the edges of the feasible space; for example, this forms the basis for the simplex algorithm for linear programming [34]. Figure 10.7 shows an example of a fitness landscape for a one-dimensional constrained optimization problem.

It can be seen that the global optimum is located at $x = 15$. However, all of the points for values $x > 15$ lie outside of the area of feasible solutions and hence, due to penalization, result in a fitness value of zero. If an algorithm finds the global optimum correctly, but due to the engineering related problems mentioned above the implementation of the solution uses $x + \varepsilon$, all solutions with a positive ε will cause a constraint violation, i.e. lead to infeasible solutions.

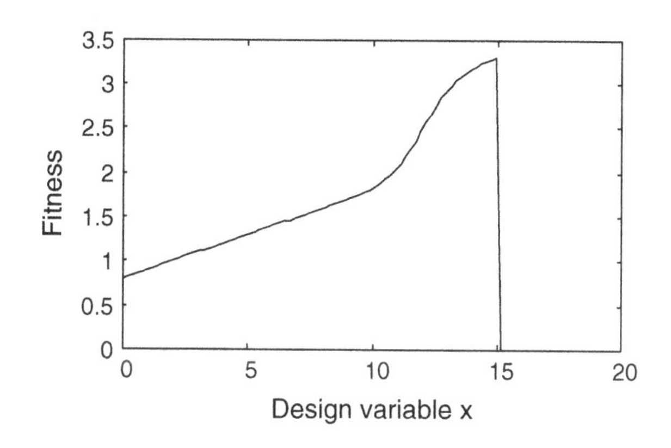

Fig. 10.7 Example of a fitness landscape for a one-dimensional constrained optimization problem

Clearly, it would be of benefit if an algorithm could finds a solution close to the global optimum but with a safety distance $d > \varepsilon$. In this case, if the realization process causes a deviation up to ε, the resulting solution would (a) still be close to the global optimum and (b) not violate any constraint.

Both problems mentioned above, i.e. narrow global optima and constraint optimization using penalty functions, are equivalent, and hence it is possible to address both issues using a single method.

This analysis led to the design of a meta-method, which can be used in conjunction with any iterative optimization algorithm. This method is presented in the next sections.

10.4 Meta-Method for Robust Optimization

Typical engineering optimization problems deal with high-dimensional and multimodal design spaces. The goal of engineering design should be to find a good enough solution that is (a) near-optimal in terms of its fitness criterion and (b) robust enough to maintain that fitness even if the manufacturing process causes slight deviations in the actual design parameter values.

Computational optimization methods have been proven to find near-optimal solutions in finite time and with a high degree of repeatability. These methods should be modified so as to learn from experience, i.e. guided away from dangerous areas during the search. However, none of the common standard computational optimization algorithms has a facility to learn during a single iteration. The closest facility would be local search, a technique often used to improve the current solution by searching the neighbourhood of the current solution. If a better solution is found, the solution itself is changed (Lamarckian learning). Alternatively, for genetic algorithms, the fitness value of the original current solution can be improved, which increases the individual's chance of being selected to generate offspring for the next generation (Baldwin learning) [15].

Figure 10.8 shows as an example a contour plot of a fitness landscape for a two-dimensional unconstrained optimization problem.

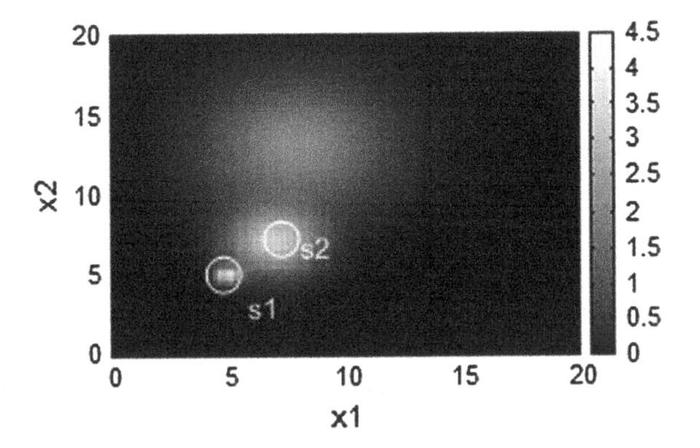

Fig. 10.8 Contour plot of a multimodal fitness landscape

As can be seen, most deviations from s1 will cause the solution to drop dramatically in fitness, whereas solution s2 is more robust in that respect. A local search here would not be of any help since the global optimum (s1) has already been found. However, if, through local probing, a measure of the deviation in fitness for the region around s1 could be established, it could be used to adjust the fitness accordingly. This would be similar to the Baldwin approach for genetic algorithms but could be used with any arbitrary optimization algorithm.

Similarly, for constraint optimization using penalty functions, this measure could be used to penalize solutions that are too close to the border with the forbidden areas. Figure 10.9 shows two solutions s1 and s2 for a two-dimensional constrained optimization problem. The grey area contains all feasible solutions; white areas are forbidden.

It can be seen that probing the areas with the radius ε around the solutions would lead to a penalization for s2, because a section of the circle is in the forbidden zone. On the other hand, s1 would not be subjected to such a penalty, because none of the points within the neighbourhood $\leq \varepsilon$ would cause a violation of a constraint.

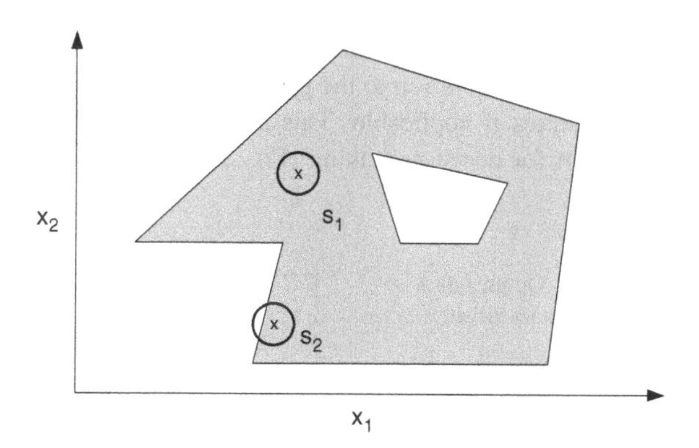

Fig. 10.9 Two different solutions in a constrained search space

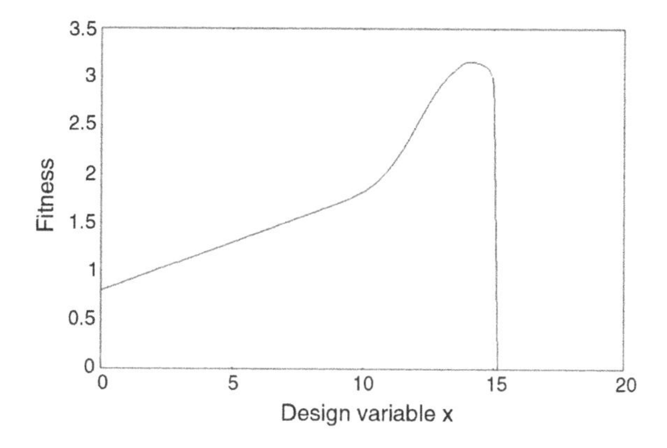

Fig. 10.10 Fitness barrier to protect solutions from dropping into the forbidden area

As can be seen above, both problems can be solved by using a measure of the quality of the solutions contained in the neighbourhood of a solution. For unconstrained optimization, this penalty would change the shape of the effective fitness landscape so that narrow peaks are penalized. For constrained optimization, it would change the shape of the fitness landscape so that points near to or on the border of a forbidden region would be lowered in fitness or, in other words, the edges would be 'rounded off' (see Fig. 10.7) and produce a fitness barrier to protect the solutions from leaving the feasible space when realized through an engineering process (Fig. 10.10).

This led to the development of the following meta-heuristic for engineering applications (Fig. 10.12). The search begins with an initial solution, respectively, with an initial population of solutions. Each time a solution is due to be evaluated by the optimization algorithm used, its fitness is evaluated first and then the neighbourhood with the radius ε is sampled with random trials. The average fitness of all of the trials is calculated and used by the host optimization algorithm instead of the fitness for the original solution. Figure 10.11 shows the optimization loop for the Baldwinean-based meta-heuristic.

It can be seen that the optimization algorithm, which could be of any type, does not receive a quality measure directly from the engineering problem. Instead, it sends its solution to the meta-heuristic, which in turn presents n slightly different versions of the solution to the problem and calculates the average fitness, including any penalties if applicable. This average fitness is then used by the optimization algorithm for decision-making (Fig. 10.12).

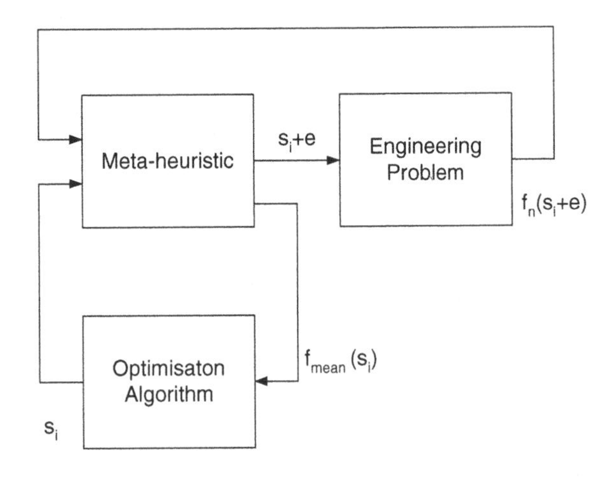

Fig. 10.11 Optimization loop for meta-heuristic

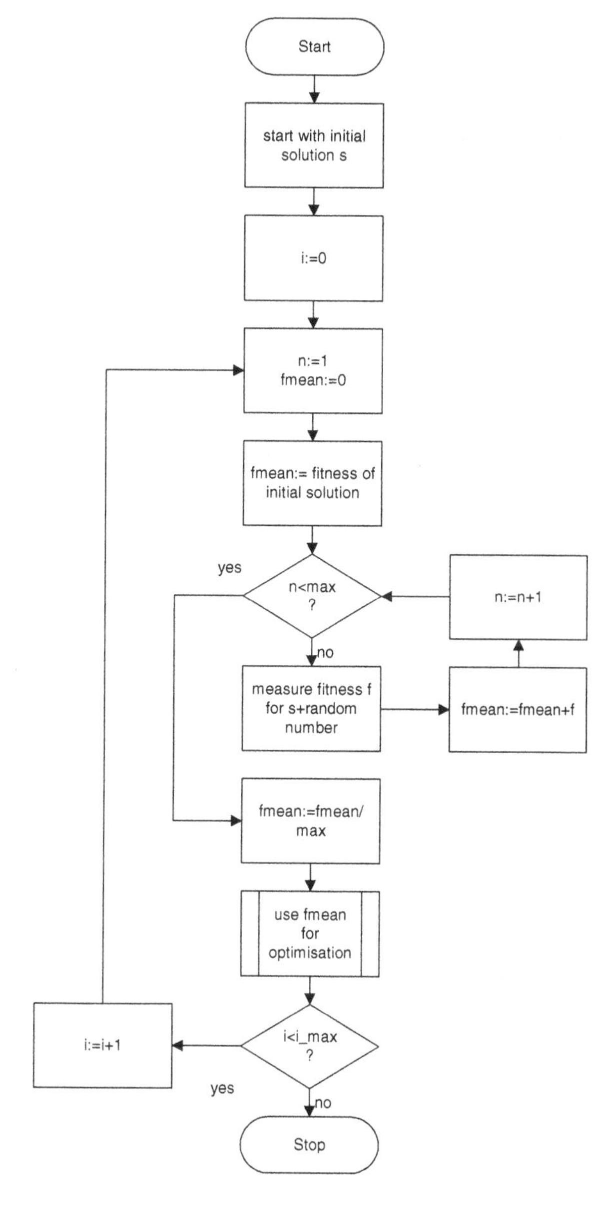

Fig. 10.12 Meta-heuristic for robust engineering

10.5 Experiments

The meta-heuristic described above can be combined with any direct search algorithm, i.e. any algorithm, that uses an optimization loop to adjust current solutions using the fitness currently observed.

For the experiments, self-adaptive stepsize search (SASS) [37] was used as the optimization algorithm.

SASS is a population-based adaptation scheme for hill climbing with a self-adaptive step size, where the temporary neighbourhood of a particle p_i is determined by the distance between itself and a randomly selected sample particle s_i of the population in each iteration. When the search is progressing, each particle is attracted by a local optimum and hence the population is clustered around a number of optima. If both, p_i and s_i are located in different clusters, p_i has the chance to escape its local optimum if it samples from a region with a higher fitness, i.e. lower costs. Towards the end of the search, most particles have reached the region of the global optimum and hence their mean distance is much smaller than in the initial population. As a result, the maximum step size $s_{\max}$ is sufficiently small to yield the global optimum.

SASS was chosen because it has only one control parameter, which is the number of particles. This eliminates the need for fine-tuning control parameters before the algorithm can be successfully applied to industrial problems [12, 38]. Hence, the experiments would not be influenced by the meta-optimization problem of tuning control parameters.

SASS was also proven to be effective and efficient for the pressure vessel problem [39]. Figure 10.13 shows the optimization loop for the pressure vessel problem, using SASS as a direct search algorithm.

For SASS, the same control parameters were used as reported by [39] to allow a fair comparison. The number of particles was hence set to 16.

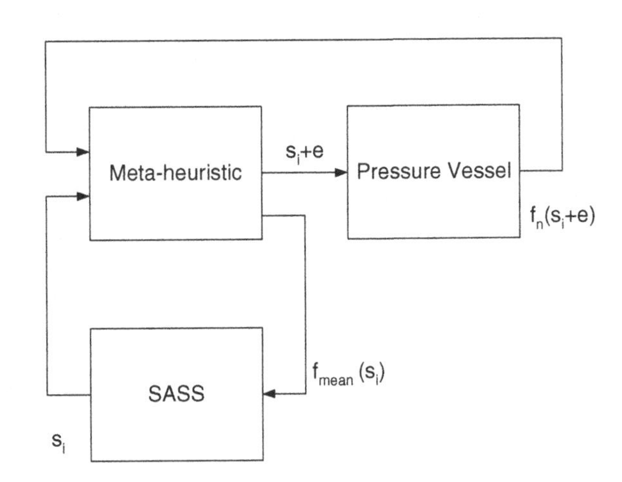

Fig. 10.13 Optimization loop for Baldwinean-based meta-heuristic

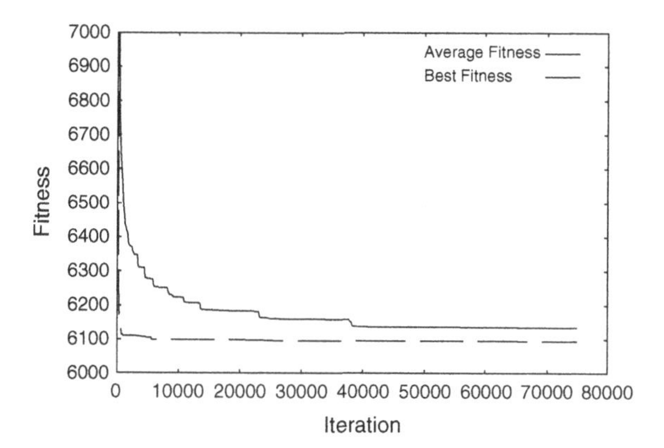

Fig. 10.14 Conversion plot for a typical Run of BMH. The *solid line* depicts the average fitness and the *dashed line* represents the best fitness in the population

Before carrying out the experiments, it was decided to limit the number of places after the decimal; for the pressure vessel problem, the design parameters $x3$ and $x4$ are continuous. For practical applications, a solution with a large number of decimal places behind the decimal point cannot be implemented, because of the accuracy of the engineering processes involved, for example ± 0.0002 in. for modern CNC machines [43]. Therefore, only four places behind the decimal point were used for the experiments.

The new method, referred to as Baldwinian-based Meta-Heuristic (BMH) has two degrees of freedom, which are the number of trials, or sample size, and the range of the samples around a solution, i.e. epsilon. These had to be determined empirically. Based on the results obtained from extensive experiments, the sample size was chosen to be 10 and epsilon was chosen to be 0.01 % of the search space dimension range.

The maximum number of iterations was chosen to be 75,000 for the experiments. After 50 runs of the algorithm, an average fitness of 6065.3908 was achieved with a standard deviation of 2.5375. Figure 10.14 shows a conversion plot of a typical run. It can be seen that the average fitness improves up to around 65,000 iterations, whereas the best fitness was already found after approximately 9,000 iterations.

The best solutions found during the 50 runs of the BMH/SASS algorithm are compared with the best solutions reported in the literature in the next section.

10.6 Evaluation

In order to evaluate the solutions obtained, a Monte Carlo simulation [46] was developed; the manufacturing process was simulated by randomly changing the optimal solutions found by different optimization algorithms. These algorithms were: Particle Swarm Optimization (PSO) [19], Hybrid Particle Swarm Branch-and-Bound (HPB) [36], Self-Adaptive Stepsize Search (SASS) [39] and the new Baldwinian-based meta-heuristic on top of SASS (BMH/SASS). Apart from the

Table 10.5 Comparison of results

	PSO	HPB	SASS	BMH/SASS
Passed (%)	42	40	40	53
Average fitness	6298.09	6298.72	6318.39	6284.24

latter method, best solutions reported in the literature were used and rounded to four decimal places behind the decimal point. The perturbation was randomly chosen from the intervals presented in Table 10.3, which represent the technical limitations for pressure vessel manufacturing [43]. A uniform distribution or random numbers was used.

As it can be seen from Table 10.5, out of the 100 pieces virtually produced, only 40 passed the quality check in the case of SASS and HPB and 42 in case of PSO. SASS was outperformed by both, PSO and HPB, in terms of pass rate and average fitness. However, when combined with BMH, the number of passes for SASS increased by 30 % and the average fitness dropped below that of PSO and HPB. This clearly shows that BMH is capable of improving the effectiveness of generic direct search algorithms for engineering applications, i.e. for applications were the actual realization of a solution differs slightly from the theoretical one because of the accuracy of the manufacturing processes involved in producing the goods in the physical world.

10.7 Conclusions

The aim of this research was to increase the robustness of engineering design solutions. Two major problems were identified; the first one is the problem of over-specifying solutions. This means, that for engineering optimization problems, the theoretical solutions have to be implemented in the physical world using manufacturing processes. As everything in the physical world, these processes suffer from noise, i.e. inaccuracies that disturb the theoretical solutions. It is also well-known that for constrained problems, optimal solutions usually exist on the borders to the feasible solution space. If the perturbation moves a solution slightly around it might enter an area of the solution space that is not allowed because one or more constrains would be violated.

The second problem is related to the narrowness of global solutions; if a global solution is located on a very narrow peak in the multi-dimensional fitness landscape, slight deviations from that location will result in a dramatic drop of fitness. In the real-world this could have catastrophic consequences for practical engineering applications. For example, if a bridge has to be built using a beam that was designed so that it has maximum strength and minimum weight, but could not be manufactured with the required accuracy, then the strength might drop below a safety level and hence the bridge could collapse. Therefore, engineers usually incorporate safety factors into their designs, which means that they move away from optimum designs. This is economically not justifiable and is in contrast to the aims of computational optimization.

To overcome these problems, a Baldwinian-based meta-heuristic (BMH) was proposed. It uses not only the fitness of a solution alone, it also probes its neighbourhood in order to estimate the goodness of the region of the solution. This meta-heuristic can be combined with any arbitrary optimization algorithm, which was demonstrated on the pressure vessel problem, where a combination of BMH and SASS was used. It was shown that BMH/SASS was able to outperform standard SASS as well as Particle Swarm Optimization (PSO) and Hybrid Particle Swarm Branch-and-Bound (HPB).

Another aspect of engineering optimization is the number of decimal places behind the decimal point. For example, a theoretical solution that relies on a design parameter to have eight decimal places behind the decimal point when measure in millimetres cannot be manufacture, even with modern CNC equipment. Therefore, a recommendation here is to use not more than four decimal places.

In conclusion, it can be said that the new method proposed in this work has the potential to find more robust solutions for engineering optimization applications.

References

1. Arnaud R, Poirion F (2014) Stochastic annealing optimization of uncertain aeroelastic system. Aerosp Sci Technol 39:456–464
2. Azad SK, Hasancebi O (2014) An elitist self-adaptive step-size search for structural design optimization. Appl Soft Comput 19:226–235
3. Azad SK, Hasancebi O (2014) Elitist self-adaptive step-size search in optimum sizing of steel structures. Int J Adv Comput Sci Appl 4(4):192–196
4. Bach H (1969) On the downhill method. Commun ACM 12(12):675–677
5. Cao YJ, Wu QH (1999) A mixed variable evolutionary programming for optimization of mechanical design. Eng Intell Syst Electr Eng Commun 7(2):77–82
6. Christensen PW, Klarbring A (2008) An introduction to structural optimization. Springer
7. Coello CAC (2002) Theoretical and numerical constraint-handling techniques used with evolutionary algorithms: a survey of the state of the art. Comput Methods Appl Mech Eng 191 (11–12):1245–1287
8. Coello CAC, Montes EM (2001) Use of dominance-based tournament selection to handle constraints in genetic algorithms. Proc ANNIE 11:177–182
9. Congedo PM, Witteveen J, Iaccarino G (2013) A simplex-based numerical framework for simple and efficient robust design optimization. Comput Optim Appl 56:231–251
10. Costanzo F (2013) Engineering mechanics: statics, 2nd edn. McGraw-Hill Companies
11. Deb K (2000) An efficient constraint handling method for genetic algorithms. Comput Methods Appl Mech Eng 186(2–4):311–338
12. Dias Junior A, da Silva Junior DC (2013) Using guiding heuristics to improve the dynamic checking of temporal properties in data dominated high-level designs. In: Proceedings of IEEE Computer Society Annual Symposium on VLSI, pp 20–25
13. Dimopoulos GG (2007) Mixed-variable engineering optimization based on evolutionary and social metaphors. Comput Methods Appl Mech Eng 196(4–6):803–817

14. Dorigo M, Gambardella L (1997) Ant colony system: a cooperative learning approach to the travelling salesman problem. IEEE Trans Evol Comput 1(1):53–66
15. El-Mihoub T, Hopgood AA, Nolle L, Battersby A (2006) Hybrid genetic algorithms: a review. Eng Lett 13(2):124–137
16. Goldberg DE (1989) Genetic algorithms in search, optimization, and machine learning. Addison-Wesley
17. Guo WA, Li WZ, Zhang Q, Wang L, Wu QD, Ren HL (2015) Biogeography-based particle swarm optimization with fuzzy elitism and its applications to constrained engineering problems. Eng Optim 46(11):1465–1484
18. Holland JH (1975) Adaptation in natural and artificial systems. University of Michigan Press
19. He S, Prempain E, Wu QH (2004) An improved particle swarm optimiser for mechanical design optimization problems. Eng Optim 36(5):585–605
20. Jamshidi R, Ghomi SMTF, Karimi B (2015) Flexible supply chain optimization with controllable lead time and shipping option. Appl Soft Comput 30:26–35
21. Jayaprakasam S, Rahim SKA, Leow CY (2015) PSOGSA-explore: a new hybrid metaheuristic approach for beam pattern optimization in collaborative beamforming. Appl Soft Comput 30:229–237
22. Jeet V, Kutanoglu E (2007) Lagrangian relaxation guided problem space search heuristics for generalized assignment problems. Eur J Oper Res 182(3):1039–1056
23. Kanagaraj G, Ponnambalam SG, Jawahar N, Nilakantan Mukund J (2015) An effective hybrid cuckoo search and genetic algorithm for constrained engineering design optimization. Eng Optim 46(10): 1331–1351
24. Kennedy J, Eberhart R (1995) Particle swarm optimization. In: Proceedings of IEEE international conference on neural networks, vol. 4, pp 1942–1948
25. Kirkpatrick S, Gelatt CD, Vecchi MP (1984) Optimization by simulated annealing: quantitative study. J Stat Phys 34(1984):975–986
26. Kitayama S, Arakawa M, Yamazaki K (2006) Penalty function approach for the mixed discrete nonlinear problems by particle swarm optimization. Struct Multidiscip Optim 32 (3):191–202
27. Li Z, Li Ze, Nguyen TT, Chen S (2015) Orthogonal chemical reaction optimization algorithm for global numerical optimization problems. Expert Syst Appl 42:3242–3252
28. Li X, Zhang G (2013) Minimum penalty for constrained evolutionary optimization. Comput Optim Appl 60(2):513–544
29. Liao T, Socha K, Montes de Oca MA, Stuetzle T, Dorigo M (2014) Ant colony optimization for mixed-variable optimization problems. IEEE Trans Evol Comput 18(4):503–518
30. Lopez RH, Ritto TG, Sampaio R, Souza de Cursi JE (2014) A new algorithm for the robust optimization of rotor-bearing systems. Eng Optim 46(8):1123–1138
31. Mahia M, Baykan ÖK, Kodaz H (2015) A new hybrid method based on particle swarm optimization, ant colony optimization and 3-opt algorithms for traveling salesman problem. Appl Soft Comput 30:484–490
32. Martínez-Soto R, Castillo O, Aguilar LT, Rodriguez A (2015) A hybrid optimization method with PSO and GA to automatically design Type-1 and Type-2 fuzzy logic controllers. Int J Mach Learn Cybernet 6:175–196
33. Metropolis N, Rosenbluth A, Rosenbluth M, Teller A, Teller E (1953) Equation of state calculations by fast computing machines. J Chem Phys 21:1087–1092
34. Murty KG (1983) *Linear programming*, John Wiley & Sons
35. Nelder JA, Mead R (1965) A simplex-method for function minimization. Comput J 7(4):308–313
36. Nema S, Goulermas J, Sparrow G, Cook P (2008) A hybrid particle swarm branch-and-bound (HPB) optimizer for mixed discrete nonlinear programming. IEEE Trans Syst Man Cybern Part A 38(6):1411–1424
37. Nolle L (2006) On a hill-climbing algorithm with adaptive step size: towards a control parameter-less black-box optimisation Algorithm. In: Reusch B (ed) Computational intelligence, theory and applications, advances in soft computing, vol 38. Springer, pp 587–595

38. Nolle L (2007) SASS applied to optimum work roll profile selection in the hot rolling of wide steel. Knowl-Based Syst 20(2):203–208
39. Nolle L, Bland JA (2012) Self-adaptive stepsize search for automatic optimal design. Knowl-Based Syst 29:75–82
40. OED (2015) Oxford English dictionary. Oxford University Press
41. Pappas M, Amba-Rao CL (1971) A direct search algorithm for automated optimum structural design. Am Inst Aeron Astron J 9(3):387–393
42. Pholdee N, Park W, Kim DK, Im Y, Bureerat S, Kwon H, Chun M (2015) Efficient hybrid evolutionary algorithm for optimization of a strip coiling process. Eng Optim 47(4), 521–532
43. Pullarcot S (2002) Practical guide to pressure vessel manufacturing. CRC Press
44. Rao SS (2009) Engineering optimization, theory and practice, 4th edn. Wiley
45. Rechenberg I (1973) Evolutionsstrategie – Optimierung technischer Systeme nach Prinzipien derbiologischen Evolution, Frommann-Holzboog
46. Rubinstein RY (1981) Simulation and the Monte Carlo method. Wiley, New York
47. Runarsson TP, Yao X (2000) Stochastic ranking for constrained evolutionary optimization. IEEE Trans Evol Comput 4(3):284–294
48. Salimi H (2015) Stochastic fractal search: a powerful metaheuristic algorithm. Knowl-Based Syst 75:1–18
49. Sandgren S (1990) Nonlinear integer and discrete programming in mechanical design optimization. J Mech Des 112:223–229
50. Sayol J, Nolle L, Schaefer G, Nakashima T (2008) Comparison of simulated annealing and SASS for parameter estimation of biochemical networks. In: Proceedings of IEEE World Congress on Computational Intelligence, 1–6 June, Hong Kong, China, pp 3568–3571
51. Shanley FR (1949) Principles of structural design for minimum weight. J Aeron Sci 16 (3):133–149
52. Standards Australia (1995) Steel plates for pressure equipment AS 1548:1995
53. Storn R, Price K (1997) Differential evolution: a simple and efficient heuristic for global optimization over continuous spaces. J Global Optim 11:341–359
54. Templeman AB (1970) Structural design or minimum cost using the method of geometric programming. ICE Proc. 46(4):459–472
55. Wang J, Yuan W, Cheng D (2015) Hybrid genetic–particle swarm algorithm: an efficient method for fast optimization of atomic clusters. Comput Theor Chem 1059:12–17
56. Xu R, Venayagamoorthy GK, Wunsch DC (2007) Modeling of gene regulatory networks with hybrid differential evolution and particle swarm optimization. Neural Netw 20(8):917–992
57. Yang HZ, Zhu Y, Lu QJ, Zhang J (2015) Dynamic reliability based design optimization of the tripod sub-structure of offshore wind turbines. Renew Energy 78:16–25
58. Zhou Y, Zhou G, Zhang J (2015) A hybrid glow worm swarm optimization algorithm to solve constrained multimodal functions optimization. Optim: J Math Progr Oper Res 64(4), 1057–1080

Author Biographies

Lars Nolle graduated from the University of Applied Science and Arts in Hanover, Germany, with a degree in Computer Science and Electronics. He obtained a PgD in Software and Systems Security and an MSc in Software Engineering from the University of Oxford as well as an MSc in Computing and a PhD in Applied Computational Intelligence from The Open University. He worked in the software industry before joining The Open University as a Research Fellow. He later became a Senior Lecturer in Computing at Nottingham Trent University and is now a Professor of Applied Computer Science and Associate Dean of Research at Jade University of Applied Sciences. His main research interests are computational optimization methods for real-world scientific and engineering applications.

Ralph Krause was born in Weimar, Germany, and studied power engineering at the University of Applied Sciences in Magdeburg. He has been working with Siemens since 1998, where he held different positions in commissioning, project management, quality management, business development and is now head of department for process, tools and training governance in medium voltage and systems.

Richard J. Cant started life as a theoretical physicist, then moved into industry, where he spent 9 years as a system designer working on computer-generated imaging for military training systems. Dr. Cant is a Senior Lecturer in Computing with the Nottingham Trent University. His areas of research interest include: Intelligent simulation and modelling, Computer graphics, Artificial intelligence, Software engineering, Integrated hardware/software design and Physical simulation.

Chapter 11
Flow-Level Packet Loss Analysis of a Markovian Bottleneck Buffer

Dieter Fiems, Stijn De Vuyst and Herwig Bruneel

Abstract Buffer overflow in intermediate network routers is the prime cause of packet loss in wired communication networks. Packet loss is usually quantified by the packet loss ratio, the fraction of packets that are lost in a buffer. While this measure captures part of the loss performance of the buffer, we show that it is insufficient to quantify the effect of loss on user-perceived quality of service for multimedia streaming applications. In this contribution, we refine the quantification of loss in two ways. First, we focus on loss of a single flow, rather than loss in a buffer. Second, we focus on the different moments of the time and number of accepted packets between losses, rather than just the mean number of accepted packets between losses (which directly relates to the packet loss ratio). The network node is modelled as a Markov-modulated $M/M/1/N$-type queueing system which is sufficiently versatile to capture the arrival correlation while keeping the analysis tractable. We illustrate our approach by some numerical examples.

Keywords Packet loss ratio · Packet flow · Queueing system

11.1 Introduction

For various multimedia streaming applications, it is well known that the packet loss ratio is an insufficient measure to assess the influence of loss on user-perceived quality of service or quality of experience. For example, Hadar et al. [15] compare various visual quality measures of an MPEG-2 encoded video stream subject to both independent and bursty packet losses and note that independent losses have a

D. Fiems (✉) · H. Bruneel
Department of Telecommunications and Information Processing,
Ghent University, St-Pietersnieuwstraat 41, 9000 Ghent, Belgium
e-mail: Dieter.Fiems@UGent.be

S. De Vuyst
Department of Industrial Systems Engineering and Product Design,
Ghent University, Technologiepark 903, 9052 Zwijnaarde, Belgium

K. Al-Begain and A. Bargiela (eds.), *Seminal Contributions to Modelling and Simulation*, Simulation Foundations, Methods and Applications,
DOI 10.1007/978-3-319-33786-9_11

more severe impact on the visual quality. Verscheure et al. [32] consider a burst-loss model (a Gilbert model) to assess the visual quality of an MPEG-2 stream. Since packet loss in wired telecommunication networks comes almost entirely from buffer overflow in intermediate network nodes, a thorough understanding of the packet loss process in a network node is essential for a reliable assessment of the quality of experience.

Let us first survey some of the literature on packet loss characteristics at a network node. The packet loss ratio is a well-studied performance measure. It is one of the main performance measures of interest of finite capacity buffer models [11, 18, 25, 26, 30]. Since it is often easier to assess the performance of infinite capacity buffer models, tail distributions [29] of the queue content of an infinite capacity buffer system are used to approximate the packet loss ratio of the corresponding finite capacity buffer system [19, 23]. For particular queueing systems, tail probabilities may even lead to exact expressions for the packet loss ratio [13].

Other loss characteristics include the probability $P(j,n)$ to have j packet losses in a block of n consecutive arrivals [1, 3, 5, 6, 8, 14], the spectrum of the packet loss process [28] and the conditional loss probability: the probability that the buffer cannot accommodate a packet, given that the preceding packet could not be accommodated as well [2, 27, 31]. In [5], the authors calculate $P(j,n)$ in a recursive way, both for $M/M/1/N$ and $IPP/M/1/N$ queueing systems. The results are compared to the corresponding probabilities as obtained under the independence assumption and it is seen that the latter are a severe underestimate of the exact values. On the other hand, if it is the case that a single packet loss causes an entire block to be invalid, then the block loss probability $1-P(0,n)$ is overestimated under the independence assumption. In [14], the same measure is derived using a purely probabilistic argument based on the classical ballot theorems. As in [5], Dán et al. [6] use a recursive scheme to calculate the probabilities $P(j,n)$ for both fixed-length packets and exponentially distributed packets. In addition to the loss measures defined above, the distribution of the number of consecutive batch losses is studied in [4] for $M^X/G/1/N$ queueing systems where batches are completely rejected if they do not fit the buffer.

Exact understanding of the packet loss characteristics is also important if one applies forward error correction (FEC) techniques. FEC may be used to mitigate the effects of packet loss and to increase the tolerable packet loss ratio but requires that packet loss is well spread in time. With FEC, only k of the n packets in a block carry essential information. The other $n-k$ packets contain redundant data that allow to recover from a limited number of lost packets in the block. In [12], the fraction of invalid blocks after FEC recovery and the mean number of consecutive invalid blocks is studied, assuming both a renewal and a Gilbert model for the loss process. Obviously, adding redundancy with FEC also increases the load of the system, which in turn adds to the number of packets lost. In [1, 3], this effect is taken into account when studying the efficiency of FEC in case of the $M/M/1/N$ loss process. In order to do so, the authors also derive the joint probability generating function of the probabilities $P(j,n)$. It is observed that redundancy degrades performance in most cases, but a sufficient amount of FEC can decrease the block loss probability if the load remains below unity. Besides the overall loss process, some authors investigate FEC performance of a selected flow in the presence of background traffic [5–7, 27]. Cidon et al. [5] study FEC performance of a selected flow in a set of Poisson flows whereas Kumazoe et al. [21] focus on FEC performance of a selected interrupted Bernoulli flow for a discrete-time queue with single-slot service times. Dan et al. [6, 7] model the selected flow as a Markovian arrival processes. The background traffic is either a Poisson (in [6]) or a Markov arrival process (in [7]) and the consecutive service times constitute sequences of exponentially, deterministically (in [6]) or Erlang (in [7]) distributed random variables. As in the FEC studies above, the main performance measure of interest are the probabilities $P(j,n)$ to have j packet losses of the selected flow in a block of n arrivals of the selected flow.

Loss performance measures of a selected flow were also studied outside the context of FEC. In [27], the authors study the loss and conditional loss probability of a selected flow for a discrete-time queueing system with single-slot service times, single server, and an uncorrelated background batch arrival process. The selected flow is assumed to be periodic. More recently, Kim et al. [20] consider flow-based performance (and loss probability) in which the complete flow may be dropped once a packet of the flow is dropped.

Flow loss measures are also the subject of the present contribution. In contrast to previous work on flow loss, we here investigate the distribution and the various moments of the time and the number of packet arrivals of a tagged flow between two consecutive losses of this flow. As loss leads to visual distortion in the context of video transmission, these characteristics provide information on the distribution of the time between these distortions and therefore on the quality of experience of the video transmission. The queueing system under investigation is an $M/M/1/N$-type queue in a Markovian environment. The queueing model at hand allows for efficient numerical performance analysis while being sufficiently versatile: the Markovian environment allows for introducing arrival correlation for the selected flow and background traffic, correlation between the selected flow and background traffic as well as for introducing phase-type service times. In addition to the loss measures above, our results also lead to expressions for the conditional loss probability and its dual notion, the conditional acceptance probability (the probability that the buffer can accommodate a packet, given that the buffer could accommodate the preceding packet). Methodologically, a transform approach is complemented by matrix techniques leading to efficient numerical algorithms for the various performance measures under investigation. This contribution is a continuation of our work [9, 10] on packet loss characteristics. In [9], similar packet loss characteristics are investigated in the context of an output buffer of an ATM packet switch with an uncorrelated arrival process. In [10], we consider packet loss for an $M/G/1$ queueing system at the buffer level.

Finally, we discuss queueing models with loss that is caused by impatience, and not by a limit on the buffer size. For such queueing models, all packets are allowed to enter the queue, but every packet has a limited queueing time. Once their maximum queueing time expires, the packets leave the queue without service. Such models are studied in [17] for generally distributed packet transmission times and generally distributed impatience times, whereas it is shown in [16] that the loss probability is minimal when the packet transmission times and impatience times are deterministic.

The remainder of this contribution is organised as follows. In the next section, we introduce the $M/M/1/N$ queueing model and derive expressions for various loss characteristics. In Sect. 11.3, the analytical model is then used to study the packet loss of the $PH+PH/PH/1/N$ queueing system by means of some numerical examples. Finally, conclusions are drawn in Sect. 11.4.

11.2 Loss Characteristics

11.2.1 Queueing Model

We consider an $M/M/1/N$-type queueing system in a Markovian environment, which means that both the instantaneous arrival and transmission rate are modulated by a continuous-time Markov chain. Let N denote the queue size (including the packet being transmitted) and let $\mathcal{K} = \{1, \ldots, K\}$ denote the state space of the Markovian environment. The state space of the resulting queueing system can be described as the product space of the possible *levels* of the queue content $\mathcal{N} = \{0, 1, \ldots, N\}$ and the *phases* in $\mathcal{K}$, resulting in $\mathcal{S} = \mathcal{N} \times \mathcal{K}$ with a total of $(N+1)K$ possible states. As the queue content can only change by one packet at a time, only transitions between states with adjacent levels occur and the generator matrix $\mathcal{Q}$ of the state transitions has a block-tridiagonal form

$$\mathcal{Q} = \begin{bmatrix} \mathcal{B}_1 & \mathcal{A}_0 & & & & \\ \mathcal{A}_2 & \mathcal{A}_1 & \mathcal{A}_0 & & & \\ & \mathcal{A}_2 & \mathcal{A}_1 & \mathcal{A}_0 & & \\ & & \ddots & \ddots & \ddots & \\ & & & \mathcal{A}_2 & \mathcal{A}_1 & \mathcal{A}_0 \\ & & & & \mathcal{A}_2 & \mathcal{A}_0 + \mathcal{A}_1 \end{bmatrix}, \tag{11.1}$$

where the blocks are of size $K \times K$ and where it is assumed that neither the arrival rate nor the transmission rate depend on the level, but only on the phase of the queue state. The matrix $\mathcal{Q}$ in (11.1) is the infinitesimal generator (see e.g. [24]) of the queueing process and its elements are not probabilities, but rather the instantaneous transition *rate* from one state in $\mathcal{S}$ to another. The diagonal elements indicate a negative rate such that the row sums of $\mathcal{Q}$ are equal to zero. The row vector p containing the equilibrium probabilities of the states in $\mathcal{S}$ satisfies the relation $p\,\mathcal{Q} = 0$. Continuous-time Markov chains with a generator matrix having a block structure as in (11.1) are known as Quasi-birth-death (QBD) processes (see also Latouche and Ramaswami [22]). Here, the arrival and transmission processes are characterised by the following matrices

- Matrix $\mathcal{A}_0$ governs the phase transitions of the underlying Markov chain when there is an arrival.
- Matrix $\mathcal{A}_1$ ($\mathcal{B}_1$) governs the phase transitions with neither arrivals nor departures when the queue is non-empty (empty).
- Matrix $\mathcal{A}_2$ governs the phase transitions with departures.

As noted before, the diagonal elements of $\mathcal{A}_1$ and $\mathcal{B}_1$ must be chosen such that the row sums of $\mathcal{A}_2 + \mathcal{A}_1 + \mathcal{A}_0$ and $\mathcal{B}_1 + \mathcal{A}_0$ are zero.

In our model, we specifically focus on the losses of a single (tagged, 't') flow of arriving packets in the presence of a background (non-tagged, 'nt') flow of packet arrivals. Therefore, we split up $\mathcal{A}_0$ into two matrices $\mathcal{A}_0^t$ and $\mathcal{A}_0^{nt}$ that contain the phase transition rates when there is an arrival that belongs to the tagged flow or to the background flow, respectively. Obviously, we then have that $\mathcal{A}_0 = \mathcal{A}_0^t + \mathcal{A}_0^{nt}$.

The class of queueing systems with QBD structure as described above includes a.o. the $PH+PH/PH/1/N$ queueing system (and also the $MAP+MAP/PH/1/N$ queueing system), the first component in the Kendall notation explicitly referring to both tagged and background arrival processes. In case of the $PH+PH/PH/1/N$ system for which we will apply our results in Sect. 11.3, the interarrival times of the tagged flow, the interarrival times of the background flow and the transmission times are all independent of each other. All three constitute a series of independent and identically distributed random variables with PH-type distribution. These PH-type distributions are characterised by their representations $(\alpha_t, \mathcal{G}_t)$, $(\alpha_{nt}, \mathcal{G}_{nt})$ and $(\alpha_s, \mathcal{G}_s)$ respectively (see e.g. Latouche and Ramaswami [22]). This then leads to the following set of transition matrices:

$$\begin{aligned}
\mathcal{A}_0^t &= (-\mathcal{G}_t \epsilon_t \alpha_t) \otimes \mathcal{I} \otimes \mathcal{I}, \\
\mathcal{A}_0^{nt} &= \mathcal{I} \otimes (-\mathcal{G}_{nt} \epsilon_{nt} \alpha_{nt}) \otimes \mathcal{I}, \\
\mathcal{A}_1 &= \mathcal{G}_t \otimes \mathcal{I} \otimes \mathcal{I} + \mathcal{I} \otimes \mathcal{G}_{nt} \otimes \mathcal{I} + \mathcal{I} \otimes \mathcal{I} \otimes \mathcal{G}_s, \\
\mathcal{B}_1 &= \mathcal{G}_t \otimes \mathcal{I} \otimes \mathcal{I} + \mathcal{I} \otimes \mathcal{G}_{nt} \otimes \mathcal{I}, \\
\mathcal{A}_2 &= \mathcal{I} \otimes \mathcal{I} \otimes (-\mathcal{G}_s \epsilon_s \alpha_s).
\end{aligned} \tag{11.2}$$

Here ϵ_i is a column vector of ones with the same number of entries as the row vector $\alpha_i (i = t, nt, s)$ and $\otimes$ is the Kronecker product as usual. The matrix $\mathcal{I}$ denotes an identity matrix of required size.

11.2.2 Uniformisation

By means of the well-known uniformisation procedure, there exists a Poisson process such that changes of the queue content and the Markovian environment only occur at the event times of this Poisson process. Let λ denote the rate of the Poisson process. This value can be chosen freely as long as it exceeds the absolute values of the diagonal elements of the matrices $\mathcal{A}_1$ and $\mathcal{B}_1$. The following $K \times K$ matrices then contain the transition probabilities of the queueing system on the event times of the Poisson process:

$$\begin{aligned} A_0^t &= \frac{1}{\lambda}\mathcal{A}_0^t, \quad A_0^{nt} = \frac{1}{\lambda}\mathcal{A}_0^{nt}, \\ A_1 &= \mathcal{I} + \frac{1}{\lambda}\mathcal{A}_1, \quad B_1 = \mathcal{I} + \frac{1}{\lambda}\mathcal{B}_1, \quad A_2 = \frac{1}{\lambda}\mathcal{A}_2. \end{aligned} \tag{11.3}$$

If we also define $A_0 = A_0^t + A_0^{nt}$, then the transition probability matrix Q of the system on Poisson events has the same block structure as $\mathcal{Q}$ in (11.1). The equilibrium probability vector π of this embedded discrete-time Markov chain satisfies $\pi Q = \pi$.

11.2.3 Analysis at Event Epochs

We now consider the state of the queueing system on the event epochs of the Poisson process. Consider the kth event epoch of the Poisson process and let U_k and S_k denote the number of packets in the queue and the state of the Markovian environment immediately after this event epoch. Clearly, U_k and S_k completely characterise the state—in the Markovian sense—of the queueing system under investigation.

To investigate packet loss characteristics, we now complement these state variables by some additional variables. One may think of these additional variables as rewards of a Markov reward process. In particular, let X_k and T_k denote the number of packet arrivals of the tagged flow since the last packet loss of the tagged flow and the time since the last loss of the tagged flow, respectively, at the kth event epoch. Clearly, when there is no loss $(U_k + A_{k+1}^t \leq N)$, we find,

$$X_{k+1} = X_k + A_{k+1}^t, \quad T_{k+1} = T_k + \Lambda_k. \tag{11.4}$$

Here A_k^t denotes the number of arrivals of the tagged flow at the kth Poisson event epoch and Λ_k denotes the time between the kth and the $(k+1)$th Poisson event epochs. Further, when there is loss of the tagged flow at the $(k+1)$th event epoch $(U_k = N$ and $A_{k+1}^t = 1)$, we have,

$$X_{k+1} = T_{k+1} = 0. \tag{11.5}$$

For $n = 0, 1, \ldots, N$ and for $j \in \mathcal{K}$ we now define the joint transform of the number of arrivals X of the tagged flow and the time T since the last loss of the tagged flow at a random Poisson event epoch in steady state, given the queue content U and the state of the environment S,

$$P_j(z,\zeta;n) = \mathrm{E}[z^X e^{-\zeta T} 1\{U=n, S=j\}] = \lim_{k\to\infty} \mathrm{E}[z^{X_k} e^{-\zeta T_k} 1\{U_k=n, S_k=j\}]. \tag{11.6}$$

In view of Eqs. (11.4) and (11.5) and by conditioning on the state of the environment and the queue content at the preceding Poisson event, we find,

$$\begin{aligned}
P_j(z,\zeta;n) &= \lim_{k\to\infty} \sum_{i\in\mathcal{K}} \sum_{m=n-1}^{n+1} \mathrm{E}[z^{X_{k+1}} e^{-\zeta T_{k+1}} 1\{U_k=m, S_k=i, U_{k+1}=n, S_{k+1}=j\}] \\
&= \lim_{k\to\infty} \sum_{i\in\mathcal{K}} \mathrm{E}[z^{X_k+1} e^{-\zeta(T_k+\Lambda_k)} 1\{U_k=n-1, S_k=i, S_{k+1}=j, A^t_{k+1}=1\}] \\
&+ \lim_{k\to\infty} \sum_{i\in\mathcal{K}} \mathrm{E}[z^{X_k} e^{-\zeta(T_k+\Lambda_k)} 1\{U_k=n-1, S_k=i, S_{k+1}=j, A^{nt}_{k+1}=1\}] \\
&+ \lim_{k\to\infty} \sum_{i\in\mathcal{K}} \mathrm{E}[z^{X_k} e^{-\zeta(T_k+\Lambda_k)} 1\{U_k=n+1, S_k=i, S_{k+1}=j, D_{k+1}=1\}] \\
&+ \lim_{k\to\infty} \sum_{i\in\mathcal{K}} \mathrm{E}[z^{X_k} e^{-\zeta(T_k+\Lambda_k)} 1\{U_k=n, S_k=i, S_{k+1}=j, A^{nt}_{k+1}=A^t_{k+1}=D_{k+1}=0\}],
\end{aligned}$$

for $0<n<N$ and where A^{nt}_{k+1} and D_{k+1} denotes the number of non-tagged arrivals and departures ar the $(k+1)$st Poisson event. The expression above further simplifies to

$$\begin{aligned}
P_j(z,\zeta;n) &= \sum_{i\in\mathcal{K}} P_i(z,\zeta;n-1)\Lambda(\zeta) z \Pr[S_{k+1}=j, A^t_{k+1}=1 | S_k=i] \\
&+ \sum_{i\in\mathcal{K}} P_i(z,\zeta;n-1)\Lambda(\zeta) \Pr[S_{k+1}=j, A^{nt}_{k+1}=1 | S_k=i] \\
&+ \sum_{i\in\mathcal{K}} P_i(z,\zeta;n+1)\Lambda(\zeta) \Pr[S_{k+1}=j, D_{k+1}=1 | S_k=i] \\
&+ \sum_{i\in\mathcal{K}} P_i(z,\zeta;n)\Lambda(\zeta) \Pr[S_{k+1}=j, A^{nt}_{k+1}=A^t_{k+1}=D_{k+1}=0 | S_k=i]
\end{aligned}$$

for $0<n<N$. Here, $\Lambda(\zeta)$ denotes the Laplace–Stieltjes transform of the time between two consecutive Poisson events

$$\Lambda(\zeta) = \frac{\lambda}{\lambda+\zeta}. \tag{11.7}$$

Similar expressions can be found for the boundary cases $n=0$ and $n=N$. Let $P(z,\zeta;n)$ denote the row vector with elements $P_i(z,\zeta;n), i\in\mathcal{K}$ and let $\pi_n = P(1,0;n)$ for $n=0,\ldots,N$. In view of the former derivations, one easily retrieves the following set of equations for the vectors $P(z,\zeta;n)$

$$\begin{aligned} P(z,\zeta;0) &= P(z,\zeta;1)\Lambda(\zeta)A_2 + P(z,\zeta;0)\Lambda(\zeta)B_1, \\ P(z,\zeta;n) &= P(z,\zeta;n-1)\Lambda(\zeta)(zA_0^t + A_0^{nt}) + P(z,\zeta;n+1)\Lambda(\zeta)A_2 + P(\zeta,z;n)\Lambda(\zeta)A_1, \\ P(z,\zeta;N) &= P(z,\zeta;N-1)\Lambda(\zeta)(zA_0^t + A_0^{nt}) + P(z,\zeta;N)\Lambda(\zeta)(A_1 + A_0^{nt}) + \pi_N A_0^t, \end{aligned} \tag{11.8}$$

for $n = 1, 2, \ldots, N-1$.

Substitution of $z = 1$ and $\zeta = 0$ in Eq. (11.8), yields the standard set of balance equations of the $M/M/1/N$ type queueing system in a Markovian environment. These equations, complemented with the normalisation condition

$$\sum_{n,k} P_k(1,0;n) = 1, \tag{11.9}$$

uniquely determine the steady-state vectors $\pi_n (n = 0, \ldots, N)$ and can be solved efficiently by means of the linear-level reduction method (see a.o. Chap. 10 of Latouche and Ramaswami [22]).

Given the vectors π_n, it is in principle possible to obtain expressions for the vectors $P(z,\zeta;n)$ $(n = 0, \ldots, N)$ from the set of Eq. (11.8). However, this requires numerous symbolic matrix manipulations and becomes infeasible for large values of either the queue size N or the size K of the state space of the Markovian environment.

Since we are mainly interested in the various moments of X and T, these cumbersome symbolic manipulations are easily avoided if one applies the moment-generating property of probability generating functions and/or Laplace–Stieltjes transforms first. We then obtain a set of equations for the various moments of X and T which can be solved efficiently. For example, for $n = 0, \ldots, N$, let t_n and x_n denote row vectors of size K with elements,

$$\begin{aligned} t_n(k) &= \mathrm{E}[T \cdot 1\{U = n, S = k\}], \\ x_n(k) &= \mathrm{E}[X \cdot 1\{U = n, S = k\}], \end{aligned} \tag{11.10}$$

respectively ($k \in \mathcal{K}$). By means of the moment-generating property of Laplace–Stieltjes transforms, we then easily obtain the following set of equations for the vectors t_n from Eqs. (11.7) and (11.8):

$$\begin{aligned} t_0 &= t_0 B_1 + t_1 A_2 + \frac{1}{\lambda}(\pi_1 A_2 + \pi_0 B_1), \\ t_n &= t_{n-1} A_0 + t_n A_1 + t_{n+1} A_2 + \frac{1}{\lambda}(\pi_{n-1} A_0 + \pi_n A_1 + \pi_{n+1} A_2), \\ t_N &= t_{N-1} A_0 + t_N (A_1 + A_0^{nt}) + \frac{1}{\lambda}(\pi_{N-1} A_0 + \pi_N (A_1 + A_0^{nt})), \end{aligned} \tag{11.11}$$

for $n = 1, 2, \ldots, N-1$. Similarly, by means of the moment-generating property of probability generating functions, we obtain the following set of equations for the vectors x_n from Eq. (11.8):

$$
\begin{aligned}
x_0 &= x_1 A_2 + x_0 B_1, \\
x_n &= x_{n-1} A_0 + x_n A_1 + x_{n+1} A_2 + \pi_{n-1} A_0^t, \\
x_N &= x_{N-1} A_0 + x_N (A_1 + A_0^{nt}) + \pi_{N-1} A_0^t,
\end{aligned} \tag{11.12}
$$

for $n = 1, 2, \ldots, N-1$. Both sets of Eqs. (11.11) and (11.12) have the following generic form:

$$
\begin{aligned}
y_0 &= y_0 C_{00} + y_1 C_{01} + c_0, \\
y_n &= y_{n-1} C_{n0} + y_n C_{n1} + y_{n+1} C_{n2} + c_n, \\
y_N &= y_{N-1} C_{N0} + y_N C_{N1} + c_N,
\end{aligned} \tag{11.13}
$$

for $n = 1, \ldots, N-1$. Here the row vectors y_n are the unknown moment vectors (conditioned on the queue content and the state of the environment) and C_{ij} and c_i are known $K \times K$ matrices and row vectors, respectively, for all i, j. Given the moments of X and/or T up to order ℓ, it is easily verified that the moment vectors of order $\ell + 1$ adhere a set of equations of the same generic form.

Algorithmic Solution The set of Eq. (11.13) can be solved efficiently by means of an adaptation of the linear-level reduction method (see Latouche and Ramaswami [22]). First, calculate the matrices D_n and row vectors d_n as follows:

$$
\begin{aligned}
D_0 &= C_{01}(I - C_{00})^{-1}, \\
d_0 &= c_0 (I - C_{00})^{-1}, \\
D_n &= C_{n2}(I - D_{n-1} C_{n0} - C_{n1})^{-1}, \\
d_n &= (d_{n-1} C_{n0} + c_n)(I - D_{n-1} C_{n0} - C_{n1})^{-1},
\end{aligned} \tag{11.14}
$$

for $n = 1, \ldots, N-1$. Recall that the matrices C_{ij} and the vectors c_i are known for all i, j (see (11.12)–(11.13)). The unknown vector y_N is then given by,

$$
y_N = d_N = (d_{N-1} C_{N0} + c_N)(I - D_{N-1} C_{N0} - C_{N1})^{-1} \tag{11.15}
$$

Finally, the vectors y_n (for $n = 0, \ldots, N-1$) can be calculated recursively as follows:

$$
y_n = y_{n+1} D_n + d_n. \tag{11.16}
$$

Operations Count Given the system of Eq. (11.13), the number of operations required to find y_0 to y_N can be calculated as follows.[1] Calculating d_0 and D_0 by

[1]Finding the $N \times M$ matrix X in $AX = B$ (A is an $N \times N$ matrix and B is an $N \times M$ matrix) by Gaussian elimination requires $\frac{1}{3}N^3 + N^2 M - \frac{1}{3}N$ multiplications and $\frac{1}{3}N^3 + (M - \frac{1}{2})N^2 - (M - \frac{1}{6})N$ additions. $N \times N$ matrix multiplication and addition requires $N^2(N-1)$ and N^2 additions and N^3 and 0 multiplications, respectively.

Gaussian elimination requires $\frac{4}{3}K^3 + K^2 - \frac{1}{3}K$ multiplications and $\frac{4}{3}K^3 - \frac{1}{2}K^2 + \frac{1}{6}K$ additions. Analogously, calculating D_n and d_n given D_{n-1} and d_{n-1} takes $\frac{7}{3}K^3 + 2K^2 - \frac{1}{3}K$ multiplications and $\frac{7}{3}K^3 + \frac{1}{2}K^2 + \frac{1}{6}K$ additions while calculating $y_N = d_N$ given D_{N-1} and d_{N-1} takes $\frac{4}{3}K^3 + 2K^2 - \frac{1}{3}K$ multiplications and $\frac{4}{3}K^3 + \frac{3}{2}K^2 + \frac{1}{6}K$ additions. Finally, calculating y_n out of y_{n+1} by (16) requires K^2 multiplications and K^2 additions. We therefore find that the total number of multiplications and additions equal,

$$\frac{7}{3}NK^3 + \frac{1}{3}K^3 + 3NK^2 + K^2 - \frac{1}{3}NK - \frac{1}{3}K$$

and

$$\frac{7}{3}NK^3 + \frac{1}{3}K^3 + \frac{3}{2}NK^2 + \frac{1}{2}K^2 + \frac{1}{6}NK + \frac{1}{6}K$$

respectively. Hence, both the number of additions and multiplications satisfy $\frac{7}{3}NK^3 + o(NK^3)$.

In addition to the solution of the set of Eq. (11.13), the calculation of the matrices C_{ij} and vectors c_{ij} requires some additional computational effort. However, only matrix addition and vector matrix multiplication is required such that overall $\frac{7}{3}NK^3 + o(NK^3)$ multiplications and additions are needed to obtain the $\ell + 1$st moment vector of X or T, given the moment vectors up to order ℓ. Moreover, calculation of the higher order moments can draw upon part of the calculations of the lower order moments. Indeed, from Eq. (11.8) it is clear that the C_{nj} do not depend on the order of the moment being calculated. Hence, the matrices D_n only have to be calculated once.

11.2.4 Performance Measures

The former results now allow for characterising the loss process in terms of the times between consecutive packet loss epochs of the tagged flow and in terms of the number of packet arrivals of the tagged flow between consecutive losses. Indeed, let $\tilde{T}$ denote the time between two consecutive packet loss epochs of the tagged flow, let $\tilde{X}$ denote the number of packets of the tagged flow that arrive in the buffer during $\tilde{T}$ and let $Q(z,\zeta)$ denote the joint transform of $\tilde{X}$ and $\tilde{T}$,

$$Q(z,\zeta) = \mathrm{E}\left[z^{\tilde{X}} e^{-\zeta \tilde{T}}\right]. \tag{11.17}$$

A packet of the tagged flow is lost at a Poisson event epoch when (i) the queue is full and (ii) there is an arrival of the tagged flow. Consider such a random loss epoch. Conditioning on the state of the environment immediately before this epoch then yields

$$Q(z,\zeta) = \Lambda(\zeta)\frac{P(z,\zeta;N)A_0^t\epsilon}{\pi_N A_0^t\epsilon}. \tag{11.18}$$

Here ϵ is a column vector with all elements equal to 1.

The moment-generating properties of probability generating functions and Laplace–Stieltjes transforms then yield expressions for the various moments of the time between consecutive losses of a tagged flow and of the number of arrivals of the tagged flow during this time. In particular, the mean time between losses of the flow μ_T and the mean number of arrivals of the tagged flow between losses μ_X are given by,

$$\mu_X = \frac{x_N A_0^t\epsilon}{\pi_N A_0^t\epsilon}, \quad \mu_T = \frac{1}{\lambda} + \frac{t_N A_0^t\epsilon}{\pi_N A_0^t\epsilon}. \tag{11.19}$$

Further, since there are on average μ_X packet arrivals (of the tagged flowPacket flow) between consecutive losses (of the tagged flow), the packet loss ratio (of the tagged flow) easily relates to μ_X as follows,

$$\text{plr} = \frac{1}{\mu_X + 1}. \tag{11.20}$$

Let the conditional loss probability (clp) denote the probability that a packet of a flow is lost given that the previous packet is lost. Since the probability that a random (tagged) loss is followed by another loss equals the probability that there are no arrivals between the tagged loss and the consecutive loss, we find that the conditional loss probability is given by

$$\text{clp} = Q(0,0) = \frac{P(0,0;N)A_0^t\epsilon}{\pi_N A_0^t\epsilon}. \tag{11.21}$$

To determine the probability $P(0,0;N)$, we notice that plugging in $\zeta = z = 0$ in Eq. (11.8) again leads to a system of equations of the generic form (11.13). In fact, the algorithm can be used to evaluate $P(\zeta,z;n)$ for any $|z| \leq 1$ and $\text{Re}(\zeta) \geq 0$ and $n = 0,\ldots,N$. This in turn allows for numerical inversion of the transforms, see Fig. 11.4 in the following section.

Similarly, let the conditional acceptance probability (cap) denote the probability that the buffer can accommodate a packet of the tagged flow, given that it could accommodate the preceding packet of the flow as well.

Consider a random arrival of the tagged flow, say arrival k. Let θ_k denote the event that the buffer can accommodate the kth packet and let $\bar{\theta}_k$ denote the event that this is not the case. We then retrieve the following expression for the conditional acceptance probability,

$$\begin{aligned} \text{cap} = \Pr[\theta_k|\theta_{k-1}] &= \frac{\Pr[\theta_k, \theta_{k-1}]}{\Pr[\theta_{k-1}]} = \frac{\Pr[\theta_{k-1}] - \Pr[\bar{\theta}_k] + \Pr[\bar{\theta}_k, \bar{\theta}_{k-1}]}{\Pr[\theta_{k-1}]} \\ &= 1 - \frac{\text{plr}}{1 - \text{plr}}(1 - \text{clp}). \end{aligned} \tag{11.22}$$

11.3 Numerical Results

To illustrate our results, we now assess the loss characteristics of a tagged flow in a $PH + PH/PH/1$ queueing system. Recall that in this case, the generator matrices are given in Eq. (11.2) in terms of the PH-type representations $(\alpha_i, \mathcal{G}_i)$ of the different distributions $(i = t, nt, s)$.

In Fig. 11.1, the mean (left) and coefficient of variation (right) of the number of packet arrivals of the tagged flow between consecutive losses of this flow are depicted versus the tagged arrival load ρ_t. Transmission and interarrival times (of the tagged flow and the background flow) are either exponentially (M/M) or hyperexponentially (H/H) distributed as indicated. For each case, three curves are depicted. The upper, middle, and lower curves correspond to queue sizes equal to 100, 50, and 20, respectively. The hyperexponential distribution under consideration has the following PH-type representation:

$$\alpha_H = \frac{1}{n}[1 \quad 1 \quad \ldots \quad 1]^T, \quad \mathcal{G}_H = -\lambda \, \text{diag}\left[\beta^0 \quad \beta^1 \quad \ldots \quad \beta^n\right], \tag{11.23}$$

where n denotes the number of phases. In Fig. 11.1, n equals 6 and β is chosen such that the coefficient of variation equals 5. The mean transmission time is set to 1 and the mean interarrival time of the background to 5/3, implying a background traffic

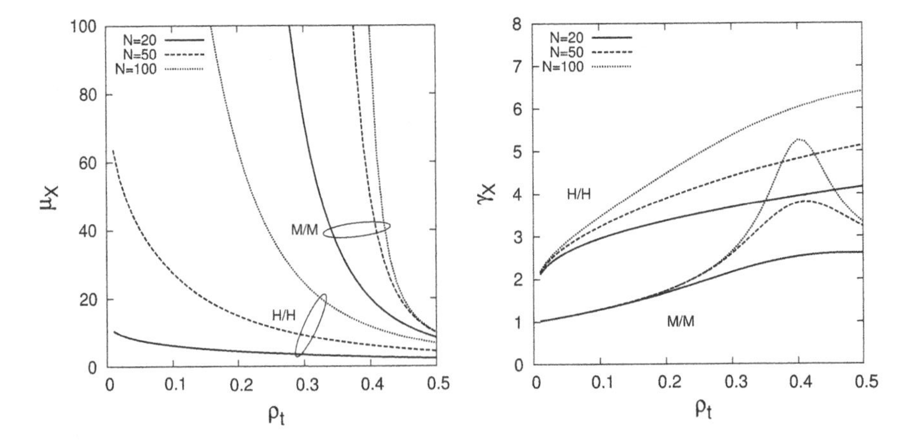

Fig. 11.1 Mean (*left*) and coefficient of variation (*right*) of the number of packet arrivals between losses of the tagged flow versus the load of the tagged flow

load of $\rho_b = 0.6$. Clearly, when the tagged load increases, μ_X decreases as expected. That is, there is more loss. Further, increasing the queue size leads to more packet arrivals between losses. Also, the M/M queueing system clearly outperforms the H/H system. This is in accordance with the frequently made observation that introducing more variance in the interarrival and transmission times leads to performance degradation. The coefficient of variation also increases for increasing tagged load at first. Again, the M/M queueing system outperforms the H/H system, the latter yielding higher coefficients of variation. For the M/M system, the coefficient of variation decreases again for $\rho_t > 0.4$ which corresponds with overloading the queue. In overload, the queue remains about full for long times which reduces variability of the queue content and therefore also of the number of packet arrivals between losses.

In Fig. 11.2, the mean time (left) and the mean number of arrivals of the tagged flow between consecutive losses of the tagged flow (right) are depicted versus the fraction ρ_t/ρ of all load that belongs to the tagged flow. In these figures, we divide the mean number of packets between losses (the mean time between losses) by the mean number of packets between losses (the mean time between losses) for low tagged load ($\rho_t = 0.02$) to clearly demonstrate the effect of the traffic mix on the loss characteristics

$$\mu_X^* = \frac{\mu_X}{\mu_X|_{\rho_t=0.02}}, \quad \mu_T^* = \frac{\mu_T}{\mu_T|_{\rho_t=0.02}}.$$

The total load is fixed and equal to $\rho = 0.8$, the mean transmission time equals 1 and the buffer can store up to $N = 30$ packets. Different interarrival time distributions are assumed as depicted. We do not specify the transmission time distribution as these curves are insensitive to the choice of distribution. Parameters of the Erlang and the hyperexponential distribution are chosen such that the coefficient of

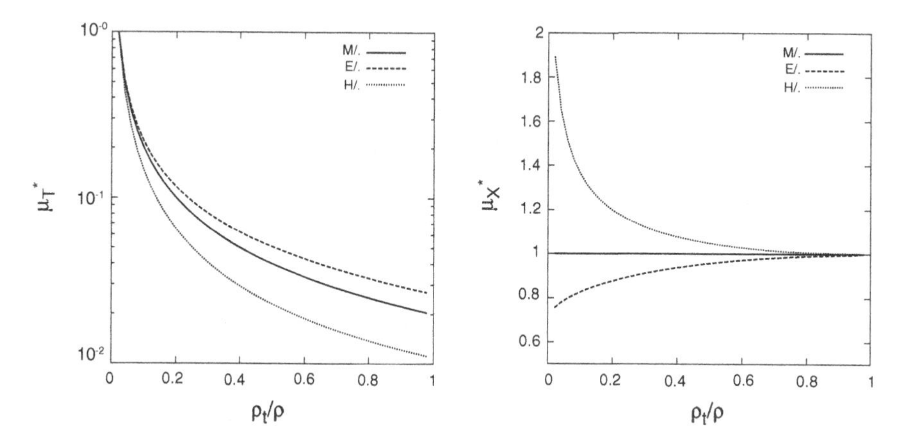

Fig. 11.2 Mean time (*left*) and mean number of arrivals of the tagged flow (*right*) between losses versus the fraction of tagged load

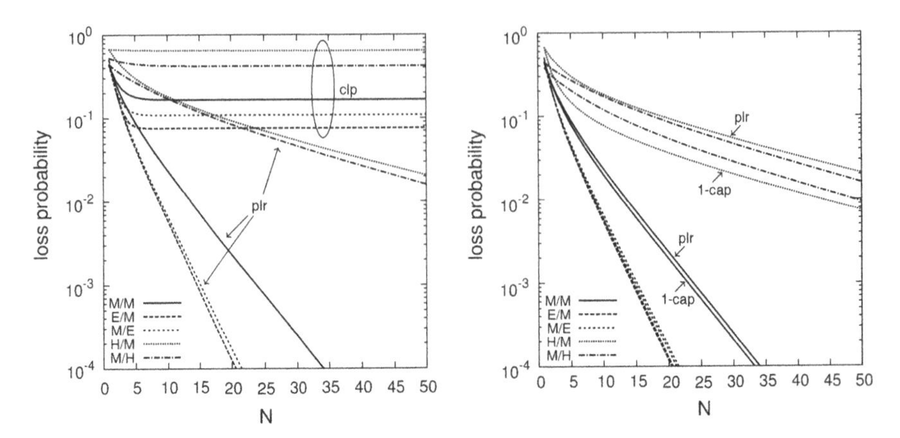

Fig. 11.3 Conditional packet loss ratio (*left*) and complement of the conditional packet acceptance ratio (*right*) versus the queue size

variation equals 1/5 and 5, respectively. Clearly, an increase of the fraction ρ_t/ρ leads to a decrease of the mean time between losses. This is what one expects since the mean interarrival time between tagged packets decreases. Further, an increase of the fraction ρ_t/ρ leads to a decrease of μ_X for hyperexponential interarrival times and to an increase for Erlang interarrival times. For exponentially distributed interarrival times, μ_X does not depend on the fraction ρ_t/ρ. This is an immediate consequence of the PASTA (Poisson Arrivals See Time Averages) property.

In Fig. 11.3 we show the conditional loss probability of the tagged flow (left) and the complement of the conditional acceptance probability of this flow versus the queue size for different interarrival time and transmission time distributions as depicted. For each interarrival time/transmission time combination, we also depict the packet loss ratio. On the clp plot (left), the lower curve and upper curve correspond to the plr and clp, respectively. On the cap plot (right), the lower curve and upper curve correspond to the complement of the cap and plr, respectively. The curves were calculated for a load of 0.8 and 90 % of the load is background traffic. As before, the mean transmission time equals 1 and the parameters of the hyper-exponential (see Eq. 11.23) and Erlang distribution are chosen such that the coefficient of variation equals 5 and 1/5, respectively.

A high clp indicates a high degree of 'clustering' in the occurrence of losses, i.e., a highly bursty loss process. As can be seen, the clp is indeed very high and more or less independent of the buffer size *N*. Nevertheless, a frequently used method of estimating the loss characteristics in a traffic stream is the 'independence' assumption, i.e., to assume that every loss occurs independently from each other. Under this assumption, we have that clp = plr, which is a severe underestimation as can be seen from Fig. 11.3. In other words, under the independence assumption the losses are well spread over time, while in reality they are much more clustered. Although one cannot use the plr to approximate the clp, one observes that the plr is a reasonable estimator for the complement of the cap.

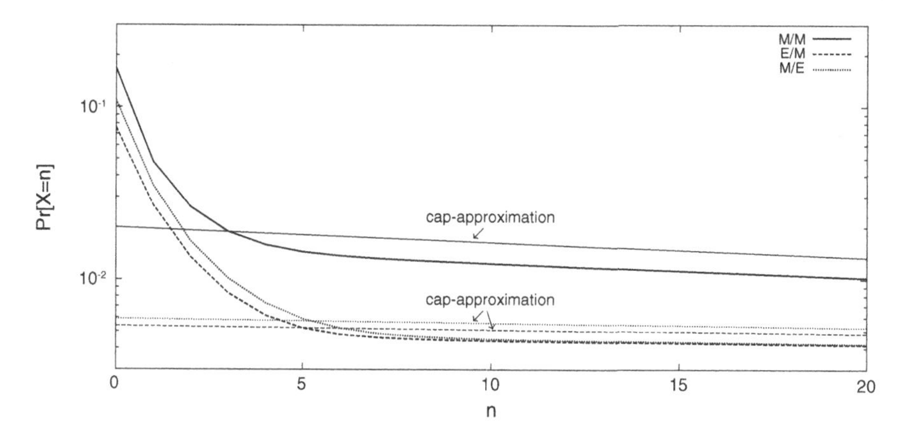

Fig. 11.4 Probability mass function of the number of arrivals between losses

Finally, in Fig. 11.4, the probability mass function of the number of arrivals in between losses is shown. Transmission and interarrival times (of the tagged flow and the background flow) are either exponentially or Erlang distributed as indicated. As before, the Erlang distributed random variables have 5 stages and the queue size is fixed at $N = 10$. For each case, two curves are depicted. The straight curve is the cap-approximation of the number of arrivals in between losses: a geometric distribution with success probability $1 -$ cap. The other curve is the distribution of the number of arrivals between losses, obtained by numerical inversion of the corresponding probability generating function. Numerical inversion (by an inverse fast Fourier transform) requires to evaluate the transform $P(\zeta, z; N)$ in $(\zeta, z) = (0, \exp(2\pi ik/K)$ for $k = 0, \ldots, K - 1$ (i denotes the imaginary unit). To this end, for each k plugging $(\zeta, z) = (0, \exp(2\pi ik/K)$ in (11.8) yields a set of equations of the generic form (11.13) which is solved by the algorithm of Sect. 11.2. In the example above, we used $K = 1024$ which is more than sufficient to suppress windowing effects.

Clearly, the cap-approximation underestimates the probability to have but a few arrivals in between losses and overestimates the probability to have many arrivals between losses. One observes that the decay rate of the distribution and of the cap-approximation are almost equal. Other numerical results not shown here confirm this observation. However, this visual observation is not confirmed by calculations: the decay rates of these curves are not equal.

11.4 Conclusions

We studied loss characteristics of $M/M/1/N$ queues in a Markovian environment. By combining a transform approach with matrix techniques, we obtained various characteristics of the loss process of a tagged flow in the presence of a background flow. We illustrated our approach by means of some numerical examples for the losses in the tagged flow in case of a $PH + PH/PH/1/N$ queue. Our results particularly demonstrate the bursty nature of the losses. We showed that a more accurate characterisation of the loss process can be retrieved efficiently by means of matrix-analytic techniques.

References

1. Ait-Hellal O, Altman E, Jean-Marie A, Kurkova I (1999) On loss probabilities in presence of redundant packets and several traffic sources. Perform Eval 36–37:485–518
2. Alouf S, Nain P, Towsley D (2001) Inferring network characteristics via moment-based estimators. Proc INFOCOM 2001:1045–1054
3. Altman E, Jean-Marie A (1998) Loss probabilities for messages with redundant packets feeding a finite buffer. IEEE J Select Areas Commun 16(5):778–787
4. Bratiychuk M, Chydzinski A (2009) On the loss process in a batch arrival queue. Appl Math Model 33(9):3565–3577
5. Cidon I, Khamisy A, Sidi M (1993) Analysis of packet loss processes in high speed networks. IEEE Trans Inf Theory IT-39(1):98–108
6. Dán G, Fodor V, Karlsson G (2006) On the effects of the packet size distribution on the packet loss process. Telecommun Syst 32(1):31–53
7. Dán G, Fodor V, Karlsson G (2006) On the effects of the packet size distribution on FEC performance. Comput Netw 50(8):1104–1129
8. Dube P, Ait-Hellal O, Altman E (2003) On loss probabilities in presence of redundant packets with random drop. Perform Eval 52(3–4):147–167
9. Fiems D, Bruneel H (2005) On higher-order packet loss characteristics. In: Proceedings of the 2005 networking and electronic commerce research conference, Riva del Garda, Italy
10. Fiems D, De Vuyst S, Wittevrongel S, Bruneel H (2009) Packet loss characteristics for M/G/1/N queueing systems. Ann Oper Res 170(1):113–131
11. Fiems D, De Vuyst S, Wittevrongel S, Bruneel H (2009) An analytic study of a scalable video buffer. Telecommun Syst 41(1):25–36
12. Frossard P (2001) FEC performance in multimedia streaming. IEEE Commun Lett 5 (3):122–124
13. Gravey A, Hébuterne G (1992) Simultaneity in discrete-time single server queues with Bernoulli inputs. Perform Eval 14:123–131
14. Gurewitz O, Sidi M, CidonI I (2000) The ballot theorem strikes again: packet loss process distribution. IEEE Trans Inf Theory 46(7):2588–2595
15. Hadar O, Huber M, Huber R, Shmueli R (2004) Quality measurements for compressed video transmitted over a lossy packet network. Opt Eng 43(2):506–520
16. Inoue Y, Takine T (2015) The *M/D/*1 + *D* queue has the minimum loss probability among M/G/1 + G queues. Oper Res Lett 43(6):629–632
17. Inoue Y, Takine T (2015) Analysis of the loss probability in the M/G/1 + G queue. Queue Syst 80(4):363–386
18. Jelenković P (1999) Subexponential loss rates in a GI/GI/1 queue with applications. Queue Syst 33:91–123
19. Kim H, Schroff N (2001) Loss probability calculations and asymptotic analysis for finite buffer multiplexers. IEEE/ACM Trans Netw 9(6):755–768
20. Kim C, Dudin S, Klimenok V (2009) The *MAP/PH/*1*/N* queue with flows of customers as a model for traffic control in telecommunication networks. Perform Eval 66:564–579
21. Kumazoe K, Kawahara K, Takine T, Oie Y (1997) Analysis of packet loss in transport layer over ATM networks. Int J Commun Syst 10:181–192

22. Latouche G, Ramaswami V (1999) Introduction to matrix analytic methods in stochastic modeling. Series on statistics and applied probability. ASA-SIAM
23. Lin GC, Suda T, Ishizaki F (2005) Loss probability for a finite buffer multiplexer with the *M/G/∞* input process. Telecommun Syst 29(3):181–197
24. Michiel H, Laevens K (1997) Teletraffic engineering in a broad-band era. Proc IEEE 85 (12):2007–2033
25. Nardelli PHJ, Kountouris M, Cardieri P, Latva-aho M (2014) Throughput optimization in wireless networks under stability and packet loss constraints. IEEE Trans Mob Comput 13 (8):1883–1895
26. Pihlsgard M (2005) Loss rate asymptotics in a GI/G/1 queue with finite buffer. Stoch Models 21(4):913–931
27. Schulzrinne H, Kurose J, Towsley D (1992) Loss correlation for queues with burtsty input streams. Proc IEEE ICC 219–224
28. Sheng H, Li S (1994) Spectral analysis of packet loss rate at a statistical multiplexer for multimedia services. IEEE/ACM Trans Network 2(1):53–65
29. Steyaert B, Xiong Y, Bruneel H (1997) An efficient solution technique for discrete-time queues fed by heterogeneous traffic. Int J Commun Syst 10:73–86
30. Takagi H (1993) Queueing analysis. Finite systems, vol 2. North-Holland
31. Takine T, Suda T, Hasegawa T (1995) Cell loss and output process analysis of a finite-buffer discrete-time ATM queueing system with correlated arrivals. IEEE Trans Commun 43:1022–1037
32. Verscheure O, Garcia X, Karlsson G, Hubaux J (1998) User-oriented QoS in packet video delivery. IEEE Netw Mag 12:12–21

Chapter 12
Fitting Methods Based on Distance Measures of Marked Markov Arrival Processes

Gábor Horváth and Miklós Telek

Abstract Approximating various real-world observations with stochastic processes is an essential modelling step in several fields of applied sciences. In this chapter, we focus on the family of Markov-modulated point processes, and propose some fitting methods. The core of these methods is the computation of the distance between elements of the model family. First, we introduce a methodology for computing the squared distance between the density functions of two phase-type (PH) distributions. Later, we generalize this methodology for computing the distance between the joint density functions of k successive inter-arrival times of Markovian arrival processes (MAPs) and marked Markovian arrival processes (MMAPs). We also discuss the distance between the autocorrelation functions of such processes. Based on these computable distances, various versions of simple fitting procedures are introduced to approximate real-world observations with the mentioned Markov modulated point processes.

Keywords Traffic modelling · Marked Markovian arrival process · Squared distance

12.1 Introduction

Phase-type (PH [17]) distributions, Markovian arrival processes (MAPs [18]), their generalizations, rational arrival processes (RAPs [1]) and their multi-type variants, MMAPs [2, 8] and MRAPs [3, 5], are versatile modelling tools in various fields of

G. Horváth (✉)
Department of Networked Systems and Services,
Budapest University of Technology and Economics, Budapest, Hungary
e-mail: ghorvath@hit.bme.hu

M. Telek
MTA-BME Information Systems Research Group, Magyar Tudósok krt. 2,
Budapest 1117, Hungary
e-mail: telek@hit.bme.hu

K. Al-Begain and A. Bargiela (eds.), *Seminal Contributions to Modelling and Simulation*, Simulation Foundations, Methods and Applications,
DOI 10.1007/978-3-319-33786-9_12

performance evaluation. If the type of an arrival event is the number of arrivals at an arrival instance, then the process is commonly referred to as batch MAP (BMAP). BMAPs and MMAPs form univocally related model classes and we consider only the second one here. PH distributions represent a dense set in the field of all positive-valued distributions [19, Theorem 8.2.8], while MAPs represent a dense class of point processes [2]. They are easy to work with: several important statistical properties can be expressed in a simple closed form, they exhibit many closeness properties, queues involving PH distributions and MAP/MMAP arrival and/or service processes can be solved efficiently, etc.

In the past decades, considerable research effort has been spent to approximate various distributions by PH distributions, to approximate various point processes by MAPs and multi-type point processes with MMAP to take the advantage of their technical simplicity. Matching and fitting methods have been developed to construct these Markovian modelling tools based on empirical measurement traces, or based on point processes like departure processes of queues, etc. However, the result produced by some of these procedures might not be ready for use immediately. There are situations when compactness (in terms of the number of states) and the Markovian representation are important.

In order to develop procedures to compress a PH, a MAP or a MMAP and/or to obtain a Markovian approximation of a non-Markovian representation, it is necessary to define distance functions which measure how "close" two stochastic processes of a model class are to each other. Since this distance function is evaluated repetitively in an optimization procedure, it must be reasonable efficient to evaluate.

In this chapter, we show that the squared distance between the density functions of two PH distributions, the joint density functions of k successive inter-arrival times of two MAPs, and the joint density and type functions of k successive inter-arrival times of two MMAPs can be expressed in a closed form. Furthermore, the squared distance between the autocorrelation functions of MAPs can be expressed in a closed form as well. Based on these results, a simple procedure is developed to approximate a non-Markovian representation by a Markovian one, and some further fitting procedure versions are discussed.

This chapter builds on [12] and extends its applicability for the multi-type version of MAPs. The rest of the chapter is organized as follows. Section 12.2 introduces the notations and the main properties of PH distributions, MAPs and RAPs used in the chapter. Section 12.3 presents how the distance between two PH distributions and Sect. 12.4 how the distance between two MAPs is calculated. The most general result of the chapter, the distance between marked Markovian arrival processes, is detailed in Sect. 12.5. The ME, RAP and MRAP approximation procedures are developed in Sect. 12.6. Finally, Sect. 12.7 demonstrates how the results are applied for the approximation of the departure process of a MAP/MAP/1 queue.

12.2 Markov Modulated Point Processes

12.2.1 Phase-Type Distributions

A phase-type (PH) distributed random variable $\mathcal{X}$ represents the time to absorption of a transient continuous time Markov chain (CTMC). PH distributions are characterized by two parameters: the generator matrix of the transient CTMC, denoted by $\boldsymbol{B}$, and the distribution of the initial state denoted by row vector β. The column vector of the rates to the absorbing state, can be computed as $\boldsymbol{b} = -\boldsymbol{B}\mathbb{1}$, where $\mathbb{1}$ denotes the column vector on ones. It means that the elements of $\boldsymbol{B}$, β and $\boldsymbol{b}$ comply with sign constraints. The off diagonal elements of $\boldsymbol{B}$ and all elements of β and $\boldsymbol{b}$ are nonnegative, while the diagonal elements of $\boldsymbol{B}$ are negative. With these notations the density function and the moments can be expressed by

$$f(x) = \beta e^{\boldsymbol{B}x}\boldsymbol{b}, \tag{12.1}$$

$$E(\mathcal{X}^i) = i!\beta(-\boldsymbol{B})^{-i}\mathbb{1}. \tag{12.2}$$

Matrix-exponential (ME, [16]) distributions are similar to PH distributions, but the above mentioned Markovian sign constraints are relaxed. This means that vector β and matrix $\boldsymbol{B}$ can be arbitrary, the only requirement is that $f(x)$ defined by (12.1) must be a valid density function.

12.2.2 Markovian Arrival Processes

A Markovian Arrival Process (MAP, [14, 18]) with N phases is given by two $N \times N$ matrices, $\boldsymbol{D}_0$ and $\boldsymbol{D}_1$. The sum $\boldsymbol{D} = \boldsymbol{D}_0 + \boldsymbol{D}_1$ is the generator of an irreducible CTMC with N states, which is the so called modulating or background process of the MAP. Consequently, each row sum of $\boldsymbol{D}$ is zero, that is

$$\boldsymbol{D}\mathbb{1} = (\boldsymbol{D}_0 + \boldsymbol{D}_1)\mathbb{1} = 0. \tag{12.3}$$

Matrix $\boldsymbol{D}_1$ contains the rates of those phase transitions which are accompanied by an arrival, and the off diagonal entries of $\boldsymbol{D}_0$ are the rates of the internal phase transitions without arrival. The sign constraints for MAPs are as follows. The off diagonal elements of $\boldsymbol{D}_0$ and all elements of $\boldsymbol{D}_1$ are nonnegative, while the diagonal elements of $\boldsymbol{D}_0$ are negative.

The phase process embedded at arrival instants plays an important role in the analysis of MAPs. This phase process is a discrete time Markov chain whose transition probability matrix is $\boldsymbol{P} = (-\boldsymbol{D}_0)^{-1}\boldsymbol{D}_1$. The stationary probability vector of the embedded process is denoted by α, it is the unique solution to linear equations $\alpha\boldsymbol{P} = \alpha, \alpha\mathbb{1} = 1$.

The joint density function of k consecutive inter-arrival times $\mathcal{X}_1, \mathcal{X}_2, \ldots \mathcal{X}_k$ is given by

$$f_k(x_1, x_2, \ldots, x_k) = \alpha e^{\boldsymbol{D_0} x_1} \boldsymbol{D_1} \cdot e^{\boldsymbol{D_0} x_2} \boldsymbol{D_1} \ldots e^{\boldsymbol{D_0} x_k} \boldsymbol{D_1} \mathbb{1}. \tag{12.4}$$

For $k = 1$, we obtain that the stationary distribution of the inter-arrival time is PH distributed with α and $\boldsymbol{D_0}$.

The lag-k autocorrelation of the inter-arrival times is matrix-geometric, and can be expressed as

$$\begin{aligned} \rho_k &= \frac{E(\mathcal{X}_1 \mathcal{X}_{k+1}) - E(\mathcal{X}_1)^2}{E(\mathcal{X}_1^2) - E(\mathcal{X}_1)^2} \\ &= \frac{\alpha(-\boldsymbol{D_0})^{-1} \boldsymbol{P}^k (-\boldsymbol{D_0})^{-1} \mathbb{1} - \alpha(-\boldsymbol{D_0})^{-1} \mathbb{1} \cdot \alpha(-\boldsymbol{D_0})^{-1} \mathbb{1}}{\sigma^2} \\ &= \frac{1}{\sigma^2} \alpha(-\boldsymbol{D_0})^{-1} (\boldsymbol{P} - \mathbb{1}\alpha)^k (-\boldsymbol{D_0})^{-1} \mathbb{1} \end{aligned} \tag{12.5}$$

for $k > 0$, and it is $\rho_0 = 1$ for $k = 0$. In (12.5), $\sigma^2 = E(\mathcal{X}_1^2) - E(\mathcal{X}_1)^2$ denotes the variance of the inter-arrival times, $E(\mathcal{X}_1)^i, i = 1, 2$ are obtained from (12.2) and $E(\mathcal{X}_1 \mathcal{X}_{k+1})$ from

$$E(\mathcal{X}_1 \mathcal{X}_{k+1}) = \int\limits_{x_1} \ldots \int\limits_{x_{k+1}} x_1 x_{k+1} f_k(x_1, x_2, \ldots, x_{k+1}) dx_{k+1} \ldots dx_1,$$

and we exploited that $\boldsymbol{P}^k - \mathbb{1}\alpha = (\boldsymbol{P} - \mathbb{1}\alpha)^k$ holds for $k > 0$ (notice, however, that it does not hold for $k = 0$).

Rational Arrival Processes (RAPs) are generalizations of MAPs, which do not obey the Markovian sign constraints. The $\boldsymbol{D_0}, \boldsymbol{D_1}$ matrices of RAPs can have arbitrary entries, the only restriction is that the joint density function must be nonnegative.

Getting rid of the Markovian sign restrictions makes RAPs more flexible than MAPs in general, but checking that a RAP is a valid stochastic process is hard (apart from the case when the transformation to a Markovian representation is successful).

12.2.3 Marked Markovian Arrival Processes

MAPs and RAPs can be extended to generate marked point processes as well. The marked versions of these processes, that are capable of representing the arrival process of multiple types of customer, are abbreviated by MMAPs and MRAPs. If there are K arrival types, MMAPs and MRAPs are characterized by $K + 1$ matrices.

The interpretation of matrix $\boldsymbol{D_0}$ is the same as in the single-class case. Matrix $\boldsymbol{D_m}$, for $m = 1, \ldots, K$, holds the phase transitions that are accompanied by a type-m arrival event. The sign constraints for MMAPs are as follows. The off diagonal elements of $\boldsymbol{D_0}$ and all elements of $\boldsymbol{D_m}$, for $m = 1, \ldots, K$ are nonnegative, while the diagonal elements of $\boldsymbol{D_0}$ are negative. Similar to MAPs $\boldsymbol{D} = \sum_{m=0}^{K} \boldsymbol{D_m}$ is the generator of an irreducible CTMC, which is the modulating process of the MMAP and consequently, $\boldsymbol{D}\mathbb{1} = 0$.

If $\mathcal{X}_k$ represents the kth inter-arrival time and $\mathcal{Y}_k$ the type of the kth customer, the stationary joint density function (including the type of the customers arrived) can be defined by

$$\begin{aligned} f_k(x_1, m_1, x_2, m_2, \ldots, x_k, m_k) \\ &= \frac{d}{dx_1}\frac{d}{dx_2}\cdots\frac{d}{dx_k} P(\mathcal{X}_1 < x_1, \mathcal{Y}_1 = m_1, \mathcal{X}_2 < x_2, \mathcal{Y}_2 = m_2, \ldots, \mathcal{X}_k < x_k, \mathcal{Y}_k = m_k) \\ &= \alpha e^{\boldsymbol{D}_0 x_1} \boldsymbol{D}_{\boldsymbol{m}_1} \cdot e^{\boldsymbol{D}_0 x_2} \boldsymbol{D}_{\boldsymbol{m}_2} \ldots e^{\boldsymbol{D}_0 x_k} \boldsymbol{D}_{\boldsymbol{m}_k} \mathbb{1}, \end{aligned} \tag{12.6}$$

where α is the stationary phase distribution at arrivals, which is the solution of $\alpha(-\boldsymbol{D_0})^{-1} \sum_{m=1}^{K} \boldsymbol{D_m} = \alpha, \alpha\mathbb{1} = 1$.

12.3 The Distance Between Two PH Distributions

In this section, we provide a simple explicit solution for the squared difference between the density functions of two PH distributions. Let us consider two PH distributed random variables, $\mathcal{A}$ and $\mathcal{B}$, with the initial probability vector, transient generator and rates to the absorbing state denoted by $(\alpha, \boldsymbol{A}, \boldsymbol{a})$ and $(\beta, \boldsymbol{B}, \boldsymbol{b})$, respectively. The squared difference between the density functions $f_{\mathcal{A}}(x)$ and $f_{\mathcal{B}}(x)$ is defined by

$$\begin{aligned} \mathcal{D}\{\mathcal{A}, \mathcal{B}\} &= \int_0^\infty (f_{\mathcal{A}}(x) - f_{\mathcal{B}}(x))^2 dx = \int_0^\infty \left(\alpha e^{\boldsymbol{A}x}\boldsymbol{a} - \beta e^{\boldsymbol{B}x}\boldsymbol{b}\right)^2 dx \\ &= L(\mathcal{A}, \mathcal{A}) - 2L(\mathcal{A}, \mathcal{B}) + L(\mathcal{B}, \mathcal{B}), \end{aligned} \tag{12.7}$$

where $L(\mathcal{B}, \mathcal{B})$ is the integral of the product of two matrix-exponentials, i.e.

$$L(\mathcal{A}, \mathcal{B}) = \int_0^\infty \alpha e^{\boldsymbol{A}x}\boldsymbol{a} \cdot \beta e^{\boldsymbol{B}x}\boldsymbol{b}\, dx. \tag{12.8}$$

The solution of this integral can be obtained by the solution of a Sylvester equation, since for compatible matrices $A, B, C, X = \int_0^\infty e^{Ax} C e^{Bx} dx$ satisfies $AX + BX + C = 0$ due to [15, Theorem 13.19]. Before applying this identity, we transpose the pdf of $\mathcal{B}$ in the integral. This step is not necessary here, but it will be essential in the subsequent sections. We get

$$L(\mathcal{A},\mathcal{B}) = \boldsymbol{b}^T \underbrace{\int_0^\infty e^{\boldsymbol{B}^T x} \beta^T \alpha e^{\boldsymbol{A} x} dx}_{\boldsymbol{Y}} \cdot \boldsymbol{a}, \tag{12.9}$$

where matrix $\boldsymbol{Y}$ is the solution to

$$-\beta^T \alpha = \boldsymbol{B}^T \boldsymbol{Y} + \boldsymbol{Y}\boldsymbol{A}. \tag{12.10}$$

This Sylvester equation has a unique solution if the eigenvalues of matrices $\boldsymbol{A}$ and $\boldsymbol{B}$ have real parts in the open left half-plane [15, Theorem 13.19], which always holds for the transient generator of PH distributions.

By applying appropriate Kronecker operations, Sylvester equations can be transformed to traditional linear equations (of form $Ax = b$, [15, Eq. 13.6]), but there are more efficient numerical solution algorithms available as well, that operate on smaller matrices by avoiding Kronecker operations. One of the fastest and most widely used direct method for solving Sylvester equations is the Hessenberg-Schur method [7], which is used by the built-in `lyap` function of Matlab at well.

12.4 Calculation of the Distance Between Two MAPs

12.4.1 *The Distance Between the Joint Density Functions of Two MAPs*

Let us consider two MAPs, $\mathcal{A} = (\boldsymbol{A}_0, \boldsymbol{A}_1)$ and $\mathcal{B} = (\boldsymbol{B}_0, \boldsymbol{B}_1)$. The squared difference of the joint density of the inter-arrival times up to lag-k is defined similar to (12.7)

$$\begin{aligned}\mathcal{D}_k\{\mathcal{A},\mathcal{B}\} = \int_0^\infty \dots \int_0^\infty \int_0^\infty & (\alpha_{\mathcal{A}} e^{\boldsymbol{A}_0 x_1} \boldsymbol{A}_1 \dots e^{\boldsymbol{A}_0 x_{k-1}} \boldsymbol{A}_1 \cdot e^{\boldsymbol{A}_0 x_k} \boldsymbol{A}_1 \mathbb{1} \\ & - \alpha_{\mathcal{B}} e^{\boldsymbol{B}_0 x_1} \boldsymbol{B}_1 \dots e^{\boldsymbol{B}_0 x_{k-1}} \boldsymbol{B}_1 \cdot e^{\boldsymbol{B}_0 x_k} \boldsymbol{B}_1 \mathbb{1})^2 dx_1 \dots dx_{k-1} dx_k,\end{aligned} \tag{12.11}$$

where $\alpha_{\mathcal{A}}$ and $\alpha_{\mathcal{B}}$ denote the stationary phase distributions of MAPs $\mathcal{A}$ and $\mathcal{B}$ at arrival instants. The square term expands to

$$\mathcal{D}_k\{\mathcal{A},\mathcal{B}\} = L_k(\mathcal{A},\mathcal{A}) - 2L_k(\mathcal{A},\mathcal{B}) + L_k(\mathcal{B},\mathcal{B}), \tag{12.12}$$

where $L_k(\mathcal{A},\mathcal{B})$ represents the integral

$$\begin{aligned} L_k(\mathcal{A},\mathcal{B}) = \int_0^\infty \ldots \int_0^\infty \int_0^\infty & \alpha_{\mathcal{A}} e^{\boldsymbol{A_0} x_1} \boldsymbol{A_1} \ldots e^{\boldsymbol{A_0} x_{k-1}} \boldsymbol{A_1} \cdot e^{\boldsymbol{A_0} x_k} \boldsymbol{A_1} \mathbb{1} \\ & \cdot \alpha_{\mathcal{B}} e^{\boldsymbol{B_0} x_1} \boldsymbol{B_1} \ldots e^{\boldsymbol{B_0} x_{k-1}} \boldsymbol{B_1} \cdot e^{\boldsymbol{B_0} x_k} \boldsymbol{B_1} \mathbb{1} dx_1 \ldots dx_{k-1} dx_k. \end{aligned} \tag{12.13}$$

The following theorem provides a procedure to evaluate this integral with the consecutive solutions of k Sylvester equations.

Theorem 12.1 *$L_k(\mathcal{A},\mathcal{B})$ can be expressed by*

$$L_k(\mathcal{A},\mathcal{B}) = \mathbb{1}^T \boldsymbol{B_1}^T \cdot \boldsymbol{Y_k} \cdot \boldsymbol{A_1} \mathbb{1}, \tag{12.14}$$

where matrix $\boldsymbol{Y_k}$ is defined recursively by Sylvester equations

$$\begin{cases} -\boldsymbol{B_1}^T \boldsymbol{Y_{k-1}} \boldsymbol{A_1} = \boldsymbol{B_0}^T \boldsymbol{Y_k} + \boldsymbol{Y_k} \boldsymbol{A_0} & \text{for } k > 1, \\ -\alpha_{\mathcal{B}}^T \alpha_{\mathcal{A}} = \boldsymbol{B_0}^T \boldsymbol{Y_1} + \boldsymbol{Y_1} \boldsymbol{A_0} & \text{for } k = 1. \end{cases} \tag{12.15}$$

Proof We start by transforming (12.13) as

$$\begin{aligned} L_k(\mathcal{A},\mathcal{B}) &= \int_0^\infty \ldots \int_0^\infty \int_0^\infty \mathbb{1}^T \boldsymbol{B_1}^T e^{\boldsymbol{B_0}^T x_k} \boldsymbol{B_1}^T e^{\boldsymbol{B_0}^T x_{k-1}} \ldots \boldsymbol{B_1}^T e^{\boldsymbol{B_0}^T x_1} \alpha_{\mathcal{B}}^T \\ &\quad \cdot \alpha_{\mathcal{A}} e^{\boldsymbol{A_0} x_1} \boldsymbol{A_1} \ldots e^{\boldsymbol{A_0} x_{k-1}} \boldsymbol{A_1} \cdot e^{\boldsymbol{A_0} x_k} \boldsymbol{A_1} \mathbb{1} dx_1 \ldots dx_{k-1} dx_k \\ &= \mathbb{1}^T \boldsymbol{B_1}^T \Bigg(\int_0^\infty \ldots \int_0^\infty \int_0^\infty e^{\boldsymbol{B_0}^T x_k} \boldsymbol{B_1}^T e^{\boldsymbol{B_0}^T x_{k-1}} \ldots \boldsymbol{B_1}^T e^{\boldsymbol{B_0}^T x_1} \alpha_{\mathcal{B}}^T \\ &\quad \cdot \alpha_{\mathcal{A}} e^{\boldsymbol{A_0} x_1} \boldsymbol{A_1} \ldots e^{\boldsymbol{A_0} x_{k-1}} \boldsymbol{A_1} \cdot e^{\boldsymbol{A_0} x_k} dx_1 \ldots dx_{k-1} dx_k \Bigg) \cdot \boldsymbol{A_1} \mathbb{1}. \end{aligned} \tag{12.16}$$

Let us denote the term in the parenthesis by $\boldsymbol{Y_k}$. For $k > 1$, separating the first and the last terms leads to the recursion

$$\begin{aligned} \boldsymbol{Y_k} &= \int_0^\infty e^{\boldsymbol{B_0}^T x_k} \cdot \boldsymbol{B_1}^T \Bigg(\int_0^\infty \ldots \int_0^\infty e^{\boldsymbol{B_0}^T x_{k-1}} \boldsymbol{B_1}^T \ldots \boldsymbol{B_1}^T e^{\boldsymbol{B_0}^T x_1} \alpha_{\mathcal{B}}^T \cdot \\ &\quad \cdot \alpha_{\mathcal{A}} e^{\boldsymbol{A_0} x_1} \boldsymbol{A_1} \ldots e^{\boldsymbol{A_0} x_{k-1}} \boldsymbol{A_1}\, dx_1 \ldots dx_{k-1} \Bigg) \boldsymbol{A_1} \cdot e^{\boldsymbol{A_0} x_k}\, dx_k \\ &= \int_0^\infty e^{\boldsymbol{B_0}^T x_k} \boldsymbol{B_1}^T \cdot \boldsymbol{Y_{k-1}} \cdot \boldsymbol{A_1} e^{\boldsymbol{A_0} x_k}\, dx_k, \end{aligned} \tag{12.17}$$

which is the solution of Sylvester equation $-\boldsymbol{B}_1^T Y_{k-1} A_1 = \boldsymbol{B}_0^T Y_k + Y_k \boldsymbol{A}_0$. The equation for $k = 1$ is obtained similarly. □

Note that the solution of (12.15) is always unique as matrices $\boldsymbol{A}_0$ and $\boldsymbol{B}_0$ are subgenerators.

12.4.2 The Distance Between the Lag Autocorrelation Functions

The squared distance between the lag autocorrelation functions of MAP $\mathcal{A}$ and $\mathcal{B}$ is computed by

$$\begin{aligned}\mathcal{D}_{\text{acf}}\{\mathcal{A},\mathcal{B}\} &= \sum_{i=0}^{\infty}(\rho_i^{(\mathcal{A})} - \rho_i^{(\mathcal{B})})^2 = \sum_{i=1}^{\infty}(\rho_i^{(\mathcal{A})} - \rho_i^{(\mathcal{B})})^2 \\ &= \sum_{i=1}^{\infty}\left(\frac{1}{\sigma_{\mathcal{A}}^2}\alpha_{\mathcal{A}}(-\boldsymbol{A}_0)^{-1}(\boldsymbol{P}_{\mathcal{A}} - \mathbb{1}\alpha_{\mathcal{A}})^i(-\boldsymbol{A}_0)^{-1}\mathbb{1}\right. \\ &\qquad \left. - \frac{1}{\sigma_{\mathcal{B}}^2}\alpha_{\mathcal{B}}(-\boldsymbol{B}_0)^{-1}(\boldsymbol{P}_{\mathcal{B}} - \mathbb{1}\alpha_{\mathcal{B}})^i(-\boldsymbol{B}_0)^{-1}\mathbb{1}\right)^2,\end{aligned} \tag{12.18}$$

where $\sigma_{\mathcal{A}}^2$ ($\sigma_{\mathcal{B}}^2$) denotes the variance of the inter-arrival times of MAP $\mathcal{A}(\mathcal{B})$, respectively, and we utilized that $\rho_i^{(\mathcal{A})} = \rho_i^{(\mathcal{B})} = 0$. Expanding the square term leads to

$$\begin{aligned}\mathcal{D}_{\text{acf}}\{\mathcal{A},\mathcal{B}\} &= \frac{1}{\sigma_{\mathcal{A}}^4}\left(M(\mathcal{A},\mathcal{A}) - m_2^{(A)2}/4\right) \\ &\quad - 2\frac{1}{\sigma_{\mathcal{A}}^2\sigma_{\mathcal{B}}^2}\left(M(\mathcal{A},\mathcal{B}) - m_2^{(A)}m_2^{(B)}/4\right) \\ &\quad + \frac{1}{\sigma_{\mathcal{B}}^4}\left(M(\mathcal{B},\mathcal{B}) - m_2^{(B)2}/4\right),\end{aligned} \tag{12.19}$$

where $m_2^{(A)}$ and $m_2^{(B)}$ denote the second moment of the inter-arrival times of MAP $\mathcal{A}$ and $\mathcal{B}$, while matrix $M(\mathcal{A},\mathcal{B})$ represents the sum

$$\begin{aligned}M(\mathcal{A},\mathcal{B}) = \sum_{i=0}^{\infty}&\alpha_{\mathcal{A}}(-\boldsymbol{A}_0)^{-1}(\boldsymbol{P}_{\mathcal{A}} - \mathbb{1}\alpha_{\mathcal{A}})^i(-\boldsymbol{A}_0)^{-1}\mathbb{1}\cdot \\ &\cdot\alpha_{\mathcal{B}}(-\boldsymbol{B}_0)^{-1}(\boldsymbol{P}_{\mathcal{B}} - \mathbb{1}\alpha_{\mathcal{B}})^i(-\boldsymbol{B}_0)^{-1}\mathbb{1}.\end{aligned} \tag{12.20}$$

The terms involving the second moments in (12.19) are necessary since the sum goes from $i = 1$ in (12.18) and it goes from $i = 0$ in (12.20). Term 0 of $M(\mathcal{A},\mathcal{B})$ equals $m_2^{(A)}/2 \cdot m_2^{(B)}/2$.

The next theorem provides the solution of matrix $M(\mathcal{A},\mathcal{B})$.

Theorem 12.2 *Matrix $M(\mathcal{A}, \mathcal{B})$ is obtained by*

$$M(\mathcal{A}, \mathcal{B}) = \alpha_{\mathcal{A}}(-\boldsymbol{A}_0)^{-1} \cdot \boldsymbol{X} \cdot (-\boldsymbol{B}_0)^{-1}\mathbb{1}, \tag{12.21}$$

where $\boldsymbol{X}$ is the unique solution to the discrete Sylvester equation

$$(\boldsymbol{P}_{\mathcal{A}} - \mathbb{1}\alpha_{\mathcal{A}}) \cdot \boldsymbol{X} \cdot (\boldsymbol{P}_{\mathcal{B}} - \mathbb{1}\alpha_{\mathcal{B}}) - \boldsymbol{X} + (-\boldsymbol{A}_0)^{-1}\mathbb{1}\alpha_{\mathcal{B}}(-\boldsymbol{B}_0)^{-1} = \boldsymbol{0}. \tag{12.22}$$

Proof Matrices $\boldsymbol{P}_{\mathcal{A}} - \mathbb{1}\alpha_{\mathcal{A}}$ and $\boldsymbol{P}_{\mathcal{B}} - \mathbb{1}\alpha_{\mathcal{B}}$ are stable (eigenvalues are inside the unit disk), since the subtraction of $\mathbb{1}\alpha_{\mathcal{A}}$ and $\mathbb{1}\alpha_{\mathcal{B}}$ removes the eigenvalue of 1 which matrices $\boldsymbol{P}_{\mathcal{A}}$ and $\boldsymbol{P}_{\mathcal{B}}$ originally had. Hence we can utilize that the solution of the sum $X = \sum_{i=0}^{\infty} A^i C B^i$ satisfies the discrete Sylvester equation $AXB - X + C = 0$. □

12.5 Calculation of the Distance Between Two MMAPs

The procedure presented in Sect. 12.4.1 can be extended for marked MAPs as well. If the number of arrival types is K, the difference between MMAPs up to lag-k is defined by

$$\begin{aligned} \mathcal{D}_k\{\mathcal{A}, \mathcal{B}\} = \sum_{m_1=1}^{K} \cdots \sum_{m_{k-1}=1}^{K} \sum_{m_k=1}^{K} \int_0^{\infty} \cdots \int_0^{\infty} \int_0^{\infty} \\ \Big(\alpha_{\mathcal{A}} e^{\boldsymbol{A}_0 x_1} \boldsymbol{A}_{m_1} \ldots e^{\boldsymbol{A}_0 x_{k-1}} \boldsymbol{A}_{m_{k-1}} \cdot e^{\boldsymbol{A}_0 x_k} \boldsymbol{A}_{m_k} \mathbb{1} \\ - \alpha_{\mathcal{B}} e^{\boldsymbol{B}_0 x_1} \boldsymbol{B}_{m_1} \ldots e^{\boldsymbol{B}_0 x_{k-1}} \boldsymbol{B}_{m_{k-1}} \cdot e^{\boldsymbol{B}_0 x_k} \boldsymbol{B}_{m_k} \mathbb{1}\Big)^2 \\ dx_1 \ldots dx_{k-1}\, dx_k, \end{aligned} \tag{12.23}$$

thus the squared distance is summed up for all combinations of arrival types up to lag-k. The expansion of the square term leads to a form similar to (12.12), but the $L_k(\mathcal{A}, \mathcal{B})$ matrices are a bit more complicated due to the different arrival types. Hence,

$$\begin{aligned} L_k(\mathcal{A}, \mathcal{B}) = \sum_{m_1=1}^{K} \cdots \sum_{m_{k-1}=1}^{K} \sum_{m_k=1}^{K} \int_0^{\infty} \cdots \int_0^{\infty} \int_0^{\infty} \\ \alpha_{\mathcal{A}} e^{\boldsymbol{A}_0 x_1} \boldsymbol{A}_{m_1} \ldots e^{\boldsymbol{A}_0 x_{k-1}} \boldsymbol{A}_{m_{k-1}} \cdot e^{\boldsymbol{A}_0 x_k} \boldsymbol{A}_{m_k} \mathbb{1} \\ \cdot \alpha_{\mathcal{B}} e^{\boldsymbol{B}_0 x_1} \boldsymbol{B}_{m_1} \ldots e^{\boldsymbol{B}_0 x_{k-1}} \boldsymbol{B}_{m_{k-1}} \cdot e^{\boldsymbol{B}_0 x_k} \boldsymbol{B}_{m_k} \mathbb{1}\, dx_1 \ldots dx_{k-1}\, dx_k. \end{aligned} \tag{12.24}$$

The multi-type (marked) counterpart of Theorem 12.1 is as follows.

Theorem 12.3 *$L_k(\mathcal{A},\mathcal{B})$ can be expressed by*

$$L_k(\mathcal{A},\mathcal{B}) = \sum_{m=1}^{K} \mathbb{1}^T \boldsymbol{B}_m^T \cdot \boldsymbol{Y}_k \cdot \boldsymbol{A}_m \mathbb{1}, \tag{12.25}$$

where matrix $\boldsymbol{Y}_k$ is the solution of the recursive Sylvester equation

$$\begin{cases} -\sum_{m=1}^{K} \boldsymbol{B}_m^T \boldsymbol{Y}_{k-1} \boldsymbol{A}_m = \boldsymbol{B}_0^T \boldsymbol{Y}_k + \boldsymbol{Y}_k \boldsymbol{A}_0 & \text{for } k > 1, \\ -\alpha_{\mathcal{B}}^T \alpha_{\mathcal{A}} = \boldsymbol{B}_0^T \boldsymbol{Y}_1 + \boldsymbol{Y}_1 \boldsymbol{A}_0 & \text{for } k = 1. \end{cases} \tag{12.26}$$

Proof The steps to prove the theorem are the same as the ones for Theorem 12.1. □

Note that the only difference between Theorems 12.1 and 12.3 is the summation in the Sylvester equation providing matrix $\boldsymbol{Y}_k$ over the different arrival types.

12.6 Application: Approximating a Non-Markovian Representation with a Markovian One

Having results for measuring the distance between two PH distributions, MAPs or MMAPs can be useful in many situations. In this section we use them as distance functions in an optimization problem. A simple procedure is developed to obtain a Markovian representation (PH/MAP/MMAP) that approximates the behaviour of a given non-Markovian (ME/RAP/MRAP) one. Two possible applications of this procedure are as follows.

- Several matching procedures produce an ME distribution, a RAP, or a MRAP which does not have a Markovian representation, or which is not even a valid stochastic process (the joint density is negative at some points). The presented procedure returns a valid Markovian representation that closely approximates the target one.
- Several performance models involve huge PH distributions, MAPs or MMAPs which make the analysis too slow and numerically demanding. With the presented procedure it is possible to compress these large models by constructing small approximate replacements that are easier (feasible) to work with.

12.6.1 Approximating an ME Distribution with a PH One

To approximate a possibly non-Markovian or too large ME distribution $\mathcal{A} = (\alpha, \boldsymbol{A}, \boldsymbol{a})$ with a Markovian PH distribution $\mathcal{B} = (\beta, \boldsymbol{B}, \boldsymbol{b})$, the following nonlinear optimization problem has to be solved:

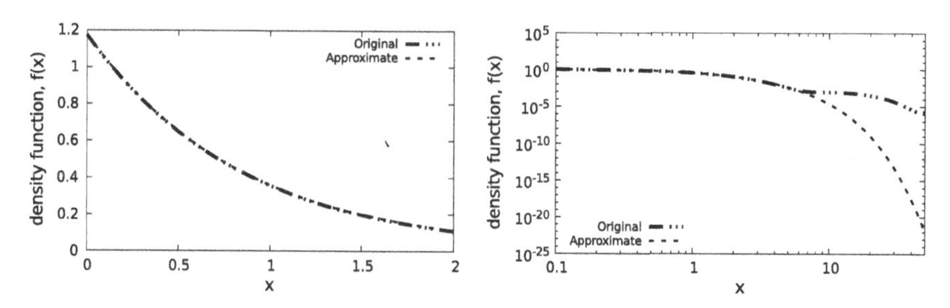

Fig. 12.1 Comparison of the density functions

$$\min_{\beta, \boldsymbol{B}} \mathcal{D}\{\mathcal{A}, \mathcal{B}\}, \tag{12.27}$$

subject to

$$\begin{aligned} \beta &\geq 0, \\ \beta \mathbb{1} &= 1, \\ \boldsymbol{B}\mathbb{1} &\leq 0, \\ [\boldsymbol{B}]_{ij} &\geq 0, \text{for } i \neq \text{j}. \end{aligned} \tag{12.28}$$

As the representation of ME distributions given by the initial probability vector and transient generator is known to be redundant, this optimization problem has infinitely many global optimum (in general).

To test this method, we calculated seven marginal moments of a measurement trace containing inter-arrival times of real data traffic,[1] and created an ME distribution by moment matching based on [22]. The procedure in [21] failed to transform the resulting ME distribution to a PH representation in an exact way, thus the method introduced here became relevant.

The nonlinear optimization problem has been solved with the built-in nonlinear solver of MATLAB, called `fmincon`. Despite of the nonlinearity of the problem and the existence of multiple optimum solutions, the optimization terminated quickly and the result turned out to be relatively independent on the initial guess.

The density functions of the original and the approximating PH distributions are depicted in Fig. 12.1. The bodies of the density functions are very close to each other, but the log–log plot reveals a drawback of the squared distance-based fitting. Namely, that the optimization sacrifices the fitting of the tiny tail densities in order to improve the distance of the body densities.

[1]We used the BC-pAug89 trace, http://ita.ee.lbl.gov/html/contrib/BC.html. While this is a fairly old trace, it is often used for testing fitting methods, it became like a benchmark.

12.6.2 Approximating a RAP with a MAP

Throughout this section, the target RAP is denoted by $\mathcal{A} = (\boldsymbol{A}_0, \boldsymbol{A}_1)$ and the approximating MAP by $\mathcal{B} = (\boldsymbol{B}_0, \boldsymbol{B}_1)$.

12.6.2.1 Obtaining Matrix B_1 Given that $\alpha_\mathcal{B}$ and B_0 are Known

Given that $\alpha_\mathcal{B}$ and $\boldsymbol{B}_0$ are already available (see later in Sect. 12.6.2.2) matrix $\boldsymbol{B}_1$ is obtained

- either to minimize $\mathcal{D}_k\{\mathcal{A},\mathcal{B}\}$ up to a given k,
- or to minimize $\mathcal{D}_{\text{acf}}\{\mathcal{A},\mathcal{B}\}$.

According to the following theorem, optimizing the squared distance of the lag-1 joint density function $\mathcal{D}_2\{\mathcal{A},\mathcal{B}\}$ is especially efficient.

Theorem 12.4 *Given that $\alpha_\mathcal{B}$ and $\boldsymbol{B}_0$ are available, matrix $\boldsymbol{B}_1$ minimizing $\mathcal{D}_2\{\mathcal{A},\mathcal{B}\}$ is the solution of the quadratic program*

$$\min_{B_1}\left\{vec\langle \boldsymbol{B}_1\rangle^T (\boldsymbol{W_{BB}} \otimes \boldsymbol{Y_{BB}}) vec\langle \boldsymbol{B}_1\rangle - 2vec\langle \boldsymbol{A}_1\rangle^T (\boldsymbol{W_{AB}} \otimes \boldsymbol{Y_{AB}}) vec\langle \boldsymbol{B}_1\rangle\right\} \quad (12.29)$$

subject to

$$\left(\boldsymbol{I} \otimes \alpha_\mathcal{B}(-\boldsymbol{B}_0)^{-1}\right) vec\langle \boldsymbol{B}_1\rangle = \alpha_A, \quad (12.30)$$

$$(\mathbb{1}^T \otimes \boldsymbol{I}) vec\langle \boldsymbol{B}_1\rangle = -\boldsymbol{B}_0\mathbb{1}. \quad (12.31)$$

Matrices $\boldsymbol{W_{AB}}, \boldsymbol{W_{BB}}, \boldsymbol{Y_{AB}}$ and $\boldsymbol{Y_{BB}}$ are the solutions to Sylvester equations

$$\boldsymbol{A}_0\boldsymbol{W_{AB}} + \boldsymbol{W_{AB}}\boldsymbol{B}_0^T = -\boldsymbol{A}_0\mathbb{1}\cdot\mathbb{1}^T\boldsymbol{B}_0^T, \quad (12.32)$$

$$\boldsymbol{B}_0\boldsymbol{W_{BB}} + \boldsymbol{W_{BB}}\boldsymbol{B}_0^T = -\boldsymbol{B}_0\mathbb{1}\cdot\mathbb{1}^T\boldsymbol{B}_0^T, \quad (12.33)$$

$$\boldsymbol{A}_0^T\boldsymbol{Y_{AB}} + \boldsymbol{Y_{AB}}\boldsymbol{B}_0 = -\alpha_A^T\cdot\alpha_\mathcal{B}, \quad (12.34)$$

$$\boldsymbol{B}_0^T\boldsymbol{Y_{BB}} + \boldsymbol{Y_{BB}}\boldsymbol{B}_0 = -\alpha_\mathcal{B}^T\cdot\alpha_\mathcal{B}. \quad (12.35)$$

Proof Let us first apply the vec$\langle\rangle$ (column stacking) operator on (12.14) at $k = 2$. Utilizing the identity $\text{vec}\langle AXB\rangle = (B^T \otimes A)\text{vec}\langle X\rangle$ for compatible matrices A, B, X and the identity $\text{vec}\langle u^T v\rangle = (v^T \otimes u^T)$ for row vectors u and v (see [20]). We get

$$\text{vec}\langle L_2(\mathcal{A},\mathcal{B})\rangle = (\mathbb{1}^T\boldsymbol{A}_0^T \otimes \mathbb{1}^T\boldsymbol{B}_0^T)\cdot\text{vec}\langle \boldsymbol{Y}_2\rangle = \text{vec}\langle \boldsymbol{B}_0\mathbb{1}\cdot\mathbb{1}^T\boldsymbol{A}_0^T\rangle^T\cdot\text{vec}\langle \boldsymbol{Y}_2\rangle. \quad (12.36)$$

Applying the vec$\langle\rangle$ operator on both sides of (12.15) and using vec$\langle AXB\rangle =$ $(B^T \otimes A)$vec$\langle X\rangle$ again leads to

$$-(\boldsymbol{I} \otimes \boldsymbol{B}_1^T \boldsymbol{Y}_1)\text{vec}\langle \boldsymbol{A}_1\rangle = (\boldsymbol{I} \otimes \boldsymbol{B}_0^T)\text{vec}\langle \boldsymbol{Y}_2\rangle + (\boldsymbol{A}_0^T \otimes \boldsymbol{I})\text{vec}\langle \boldsymbol{Y}_2\rangle, \tag{12.37}$$

from which vec$\langle \boldsymbol{Y}_2\rangle$ is expressed by

$$\text{vec}\langle \boldsymbol{Y}_2\rangle = (-\boldsymbol{A}_0^T \oplus \boldsymbol{B}_0^T)^{-1}(\boldsymbol{I} \otimes \boldsymbol{B}_1^T)(\boldsymbol{I} \otimes \boldsymbol{Y}_{\boldsymbol{AB}})\text{vec}\langle \boldsymbol{A}_1\rangle, \tag{12.38}$$

since $\boldsymbol{Y}_1 = \boldsymbol{Y}_{\boldsymbol{AB}}$. Thus we have

$$\text{vec}\langle L_2(\mathcal{A}, \mathcal{B})\rangle = \underbrace{\text{vec}\langle \boldsymbol{B}_0 \mathbb{1} \cdot \mathbb{1}^T \boldsymbol{A}_0^T\rangle^T (-\boldsymbol{A}_0^T \oplus \boldsymbol{B}_0^T)^{-1}}_{\text{vec}\langle \boldsymbol{W}_{\boldsymbol{AB}}\rangle^T} (\boldsymbol{I} \otimes \boldsymbol{B}_1^T)(\boldsymbol{I} \otimes \boldsymbol{Y}_{\boldsymbol{AB}})\text{vec}\langle \boldsymbol{A}_1\rangle, \tag{12.39}$$

where we recognized that the transpose of vec$\langle \boldsymbol{W}_{\boldsymbol{AB}}\rangle$ expressed from (12.32) matches the first two terms of the expression. Using the identities of the vec$\langle\rangle$ operator yields

$$\text{vec}\langle \boldsymbol{W}_{\boldsymbol{AB}}\rangle^T (\boldsymbol{I} \otimes \boldsymbol{B}_1^T) = \text{vec}\langle \boldsymbol{B}_1^T \boldsymbol{W}_{\boldsymbol{AB}}\rangle^T = \text{vec}\langle \boldsymbol{B}_1\rangle^T (\boldsymbol{W}_{\boldsymbol{AB}} \otimes \boldsymbol{I}). \tag{12.40}$$

Finally, putting together (12.39) and (12.40) gives

$$\text{vec}\langle L_2(\mathcal{A}, \mathcal{B})\rangle = \text{vec}\langle \boldsymbol{B}_1\rangle^T (\boldsymbol{W}_{\boldsymbol{AB}} \otimes \boldsymbol{Y}_{\boldsymbol{AB}})\text{vec}\langle \boldsymbol{A}_1\rangle. \tag{12.41}$$

From the components of $\mathcal{D}_2\{\mathcal{A}, \mathcal{B}\}$ (see 12.12) $L_2(\mathcal{A}, \mathcal{A})$ plays no role in the optimization as it does not depend on $\boldsymbol{B}_1$, the term $L_2(\mathcal{A}, \mathcal{B})$ yields the linear term in (12.29) according to (12.41), and $L_2(\mathcal{B}, \mathcal{B})$ introduces the quadratic term, based on (12.41) after replacing $\mathcal{A}$ by $\mathcal{B}$.

According to the first constraint (12.30) and the second constraint (12.31) the solution must satisfy $\alpha_{\mathcal{B}}(-\boldsymbol{B}_0)^{-1}\boldsymbol{B}_1 = \alpha_{\mathcal{B}}$ and $\boldsymbol{B}_1 \mathbb{1} = -\boldsymbol{B}_0 \mathbb{1}$, respectively. □

Theorem 12.5 *Matrix $\boldsymbol{W}_{\boldsymbol{BB}} \otimes \boldsymbol{Y}_{\boldsymbol{BB}}$ is positive definite, thus the quadratic optimization problem of Theorem* 12.4 *is convex.*

Proof If $\boldsymbol{W}_{\boldsymbol{BB}}$ and $\boldsymbol{Y}_{\boldsymbol{BB}}$ are positive definite, then their Kronecker product is positive definite as well. First we show that matrix $\boldsymbol{Y}_{\boldsymbol{BB}}$ is positive definite, thus $z\boldsymbol{Y}_{\boldsymbol{BB}}z^T > 0$ holds for any nonzero row vector z. Since $\boldsymbol{Y}_{\boldsymbol{BB}}$ is the solution of a Sylvester equation, we have that $\boldsymbol{Y}_{\boldsymbol{BB}} = \int_0^\infty e^{\boldsymbol{B}_0^T x} \alpha_{\mathcal{B}}^T \cdot \alpha_{\mathcal{B}} e^{\boldsymbol{B}_0 x}\, dx$. Hence

$$z\boldsymbol{Y}_{\boldsymbol{BB}}z^T = \int_0^\infty z e^{\boldsymbol{B}_0{}^T x} \alpha_{\mathcal{B}}^T \cdot \alpha_{\mathcal{B}} e^{\boldsymbol{B}_0 x} z^T\, dx = \int_0^\infty \left(\alpha_{\mathcal{B}} e^{\boldsymbol{B}_0 x} z^T\right)^2 dx, \tag{12.42}$$

which cannot be negative, furthermore, apart from a finite number of x values $\alpha_{\mathcal{B}} e^{\boldsymbol{B}_0 x} z^T$ cannot be zero either. Thus, the integral is always strictly positive.

The positive definiteness of matrix $\boldsymbol{W_{BB}}$ can be proven similarly. □

Being able to formalize the optimization of $\mathcal{D}_2\{\mathcal{A},\mathcal{B}\}$ as a quadratic programming problem means that obtaining the optimal matrix $\boldsymbol{B_1}$ is efficient: it is fast, and there is a single optimum which is always found.

If we intend to take higher lag joint density differences also into account, the objective function is $\mathcal{D}_k\{\mathcal{A},\mathcal{B}\}$, which is not quadratic for $k > 2$. However, our numerical experience indicates that the built-in nonlinear optimization tool in MATLAB, called `fmincon` is able to return the solution matrix $\boldsymbol{B_1}$ quickly, independent of the initial point of the optimization. We have a strong suspicion that the returned solution is the global optimum, however, we cannot prove the convexity of the objective function formally.

It is also possible to use $\mathcal{D}_{acf}\{\mathcal{A},\mathcal{B}\}$ as the objective function of the optimization problem, when looking for matrix $\boldsymbol{B_1}$ that minimizes the squared difference of the autocorrelation function. We found that `fmincon` is rather prone to the initial point in this case. Repeated running with different random initial points was required to obtain the best solution.

12.6.2.2 Approximating a RAP

The proposed procedure consists of two steps:

1. obtaining the phase-type (PH) representation of the stationary inter-arrival times, that provides vector $\alpha_{\mathcal{B}}$ and matrix $\boldsymbol{B_0}$;
2. obtaining the optimal $\boldsymbol{B_1}$ matrix which minimizes the distance of the correlation structure with the target RAP.

Section 12.6.2.1 describes step 2. For step 1, any phase-type fitting method can be applied. To solve this problem [6] develops a moment matching method that returns a hyper-exponential distribution of order N based on $2N - 1$ moments, if it is possible. Another solution published in [13] is based on a hyper-Erlang distribution, which always succeeds if an appropriately large Erlang order is chosen.

Our method of choice, however, is a slight modification of [10], which is the generalization of the former two. It constructs PH distributions from feedback Erlang blocks (FEBs, see Fig. 12.2), where each FEB implements an eigenvalue of the target distribution. An FEB of a single state represents areal eigenvalue. With FEBs it is possible to represent complex eigenvalues as well, as opposed to the previously mentioned methods that operate on hyper-exponential and hyper-Erlang distributions. The original method in [10] puts the FEBs in a row (as it is in Fig. 12.3), which is not appropriate for our goals, since there is only a single absorbing state, implying that matrix $\boldsymbol{B_1}$ can have only a single nonzero row, thus no correlation can be realized. However, the original method can be modified in a straight forward way to return a hyper-FEB structure (as it is in Fig. 12.4). A key

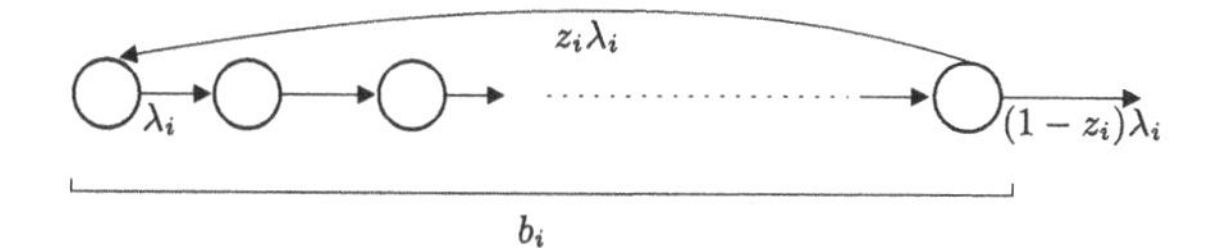

Fig. 12.2 Structure of a feedback Erlang block

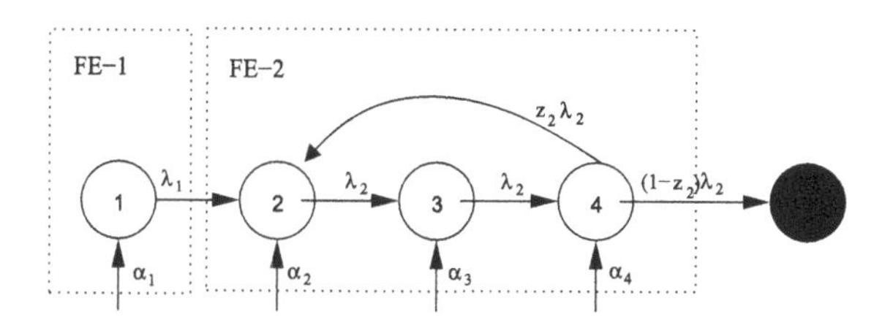

Fig. 12.3 PH distribution composed of serial FEBs

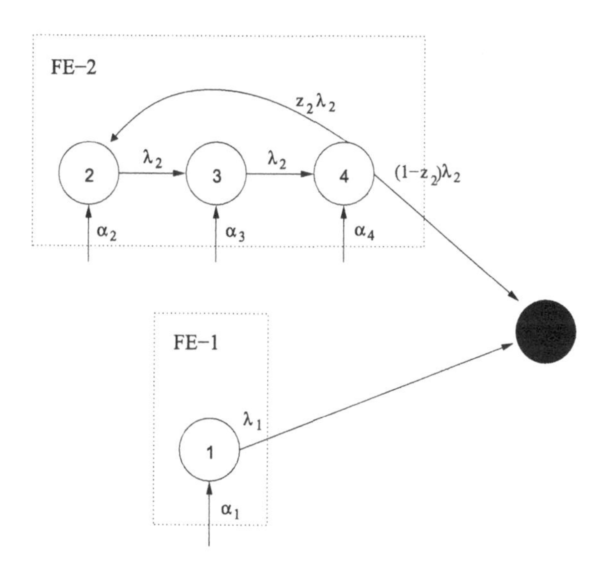

Fig. 12.4 PH distribution composed of parallel FEBs

step of [10] is the solution of a polynomial system of equations, which can have several solutions, providing several valid $\alpha_{\mathcal{B}}, \boldsymbol{B_0}$ pairs. Our RAP approximation procedure performs the optimization of matrix $\boldsymbol{B_1}$ with all of these solutions, and picks the best one among them.

12.6.2.3 Numerical Examples

In the first numerical example we extract seven marginal moments and nine lag-1 joint moments from the measurement trace used in Sect. 12.6.1, and create a RAP of order 4 with the method published in [21]. The obtained matrices are as follows:

$$A_0 = \begin{bmatrix} -0.579 & -0.402 & -0.364 & -0.348 \\ -0.368 & -0.205 & -0.315 & -0.36 \\ 1.32 & -0.845 & 0.701 & 1.13 \\ -1.7 & 0.3 & -1.14 & -1.52 \end{bmatrix},$$
$$A_1 = \begin{bmatrix} 0.576 & 0.262 & 0.41 & 0.446 \\ 0.168 & 0.501 & 0.313 & 0.266 \\ 0.29 & -1.69 & -0.598 & -0.302 \\ 0.292 & 1.94 & 1.03 & 0.786 \end{bmatrix}.$$

The RAP characterized by $\mathcal{A} = (A_0, A_1)$ is, however, not a valid stochastic process as the joint density given by (12.4) is negative since $f_2(0.5, 8) = -0.000357$. This RAP is the target of our approximation in this section.

Let us now construct a MAP $\mathcal{B}^{(1)} = (B_0^{(1)}, B_1^{(1)})$ which minimizes the squared distance of the lag-1 joint density with $\mathcal{A}$. The distribution of the inter-arrival times, characterized by $\alpha_{\mathcal{B}}, B_0^{(1)}$ are obtained by the modified moment matching method of [10], and matrix $B_1^{(1)}$ has been determined by the quadratic program provided by Theorem 12.4. The matrices of the MAP are

$$B_0^{(1)} = \begin{bmatrix} -0.074 & 0 & 0 & 0 & 0 \\ 0 & -0.27 & 0.27 & 0 & 0 \\ 0 & 0 & -0.27 & 0.27 & 0 \\ 0 & 0 & 0 & -0.27 & 0 \\ 0 & 0 & 0 & 0 & -1.2 \end{bmatrix},$$
$$B_1^{(1)} = \begin{bmatrix} 0.0065 & 0.024 & 0 & 5.5 \cdot 10^{-8} & 0.044 \\ 0 & 0 & 0 & 0 & 0 \\ 0 & 0 & 0 & 0 & 0 \\ 0.017 & 0.086 & 0 & 0 & 0.17 \\ 0 & 0.012 & 0 & 0 & 1.2 \end{bmatrix},$$

and the squared distance in the lag-1 joint pdf is $\mathcal{D}_2\{\mathcal{A}, \mathcal{B}^{(1)}\} = 0.000105$. The quadratic program has been solved by MATLAB is less than a second. Next, we repeat the same procedure, but instead of focusing on the lag-1 distance, we optimize on the squared distance of the joint pdf up to lag-10. This cannot be formalized as a quadratic program any more, but the optimization is still fast, lasting only 1–2 s. In this case, the hyper-exponential distribution provided the best results ($\mathcal{D}_{11}\{\mathcal{A}, \mathcal{B}^{(10)}\} = 4.37 \times 10^{-5}$). The matrices are

$$B_0^{(10)} = \begin{bmatrix} -0.0519 & 0 & 0 \\ 0 & -0.151 & 0 \\ 0 & 0 & -1.24 \end{bmatrix},$$

$$B_1^{(10)} = \begin{bmatrix} 10^{-6} & 0.0519 & 10^{-6} \\ 10^{-6} & 0.151 & 0.000465 \\ 0.000129 & 10^{-6} & 1.24 \end{bmatrix}.$$

To evaluate the quality of the approximation Fig. 12.5 compares the marginal density functions of $\mathcal{A}, \mathcal{B}^{(1)}$ and $\mathcal{B}^{(10)}$. The plots are close to each other, the approximation is relatively accurate. To demonstrate that the lag-1 joint densities are also accurate, Fig. 12.6 depicts them at $x_2 = 0.5, 1$ and 1.5.

In the next experiment the objective is the squared distance of the lag-k autocorrelation function. As before, the input RAP is $\mathcal{A}$, but now the approximation procedure has to minimize $\mathcal{D}_{\text{acf}}\{\mathcal{A}, \mathcal{B}^{(\rho)}\}$ which is given in a closed form by (12.19) and Theorem 12.2. According to our experience, the result of the

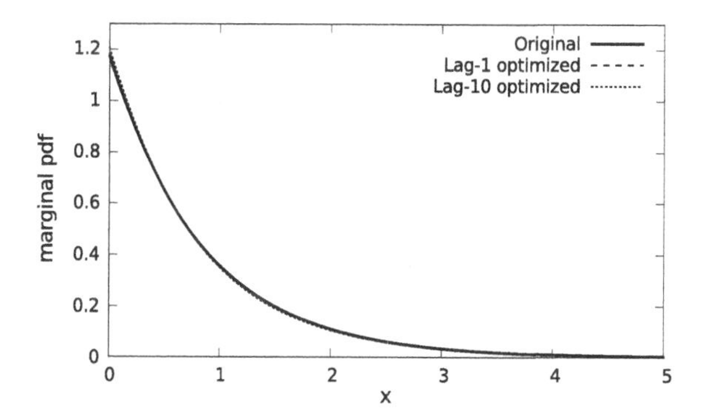

Fig. 12.5 Comparison of the density functions of the marginal distribution

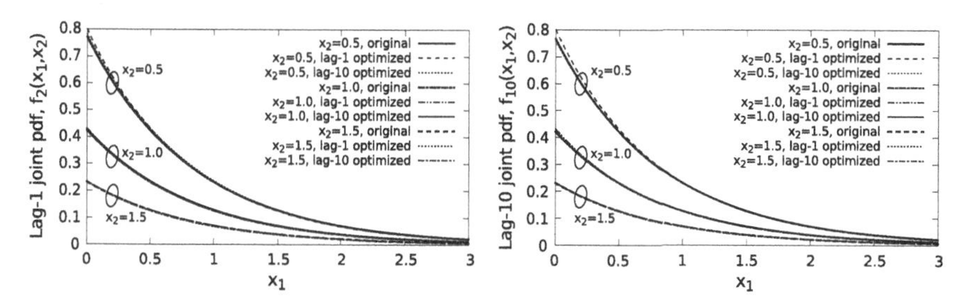

Fig. 12.6 Comparison of the lag-1 joint density functions

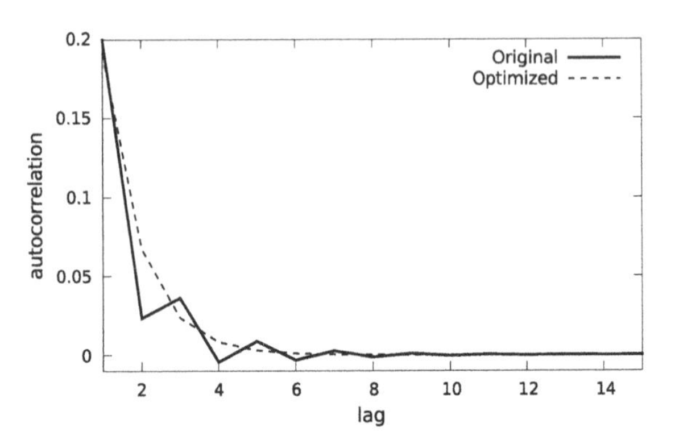

Fig. 12.7 Comparison of the autocorrelation functions

optimization is rather prone to the initial point. The best result from 10 trials is given by matrices

$$
\boldsymbol{B}_0^{(\rho)} = \begin{bmatrix} -0.0851 & 0.0851 & 0 & 0 & 0 & 0 \\ 0 & -0.0851 & 0 & 0 & 0 & 0 \\ 0 & 0 & -0.267 & 0.267 & 0 & 0 \\ 0 & 0 & 0 & -0.267 & 0.267 & 0 \\ 0 & 0 & 0 & 0 & -0.267 & 0 \\ 0 & 0 & 0 & 0 & 0 & -1.2 \end{bmatrix},
$$

$$
\boldsymbol{B}_1^{(\rho)} = \begin{bmatrix} 0 & 0 & 0 & 0 & 0 & 0 \\ 0 & 0 & 0.0485 & 0 & 0 & 0.0366 \\ 0 & 0 & 0 & 0 & 0 & 0 \\ 0 & 0 & 0 & 0 & 0 & 0 \\ 0 & 0 & 0.0965 & 0 & 0 & 0.1705 \\ 0.0004 & 0 & 0.0117 & 0 & 0 & 1.1885 \end{bmatrix}.
$$

and the corresponding autocorrelation function is depicted in Fig. 12.7. The squared distance between the autocorrelation functions is $\mathcal{D}_{\text{acf}}\{\mathcal{A}, \mathcal{B}^{(\rho)}\} = 0.00237$.

12.6.3 Approximating a MRAP with a MMAP

If the number of arrival types K is greater than 1, the idea presented in Sect. 12.6.2.1 to formalize the approximation as a quadratic optimization problem cannot be applied. To obtain the matrices $\boldsymbol{B}_m, m = 1, \ldots, K$, assuming that $\boldsymbol{B}_0$ is known, the following nonlinear optimization problem has to be solved:

$$\min_{B_1,\ldots,B_K} \mathcal{D}_k\{\mathcal{A},\mathcal{B}\}, \tag{12.43}$$

where $\mathcal{D}_k\{\mathcal{A},\mathcal{B}\}$ is the multi-type distance up to lag-k as defined by (12.23). The inequality and equality constraints (to ensure that $\sum_{m=0}^{K} \boldsymbol{B}_m \mathbb{1} = 0$ and that the stationary phase distribution at arrivals is β) are

$$\begin{aligned} \boldsymbol{B}_m &\geq \boldsymbol{0}, \text{for } m = 1,\ldots,K, \\ \sum_{m=1}^{K} \boldsymbol{B}_m \mathbb{1} &= -\boldsymbol{B}_0 \mathbb{1}, \\ \beta(-\boldsymbol{B}_0)^{-1} \sum_{m=1}^{K} \boldsymbol{B}_m &= \beta. \end{aligned} \tag{12.44}$$

While many nonlinear programming problems are difficult to solve, we had a positive numerical experience with this one. The `fmincon` method of MATLAB managed to terminate in a couple of seconds returning a valid solution that is relatively independent on the initial guess.

To demonstrate the usefulness of this procedure, the numerical example in paper [11] is revisited. In that example two MMAP[K]/PH[K]/1-FCFS queues are considered in a tandem setting, and the performance measures of the second queue are analyzed. The traffic on the link between the queues is approximated by a MMAP, which is obtained by matching the lag-1 joint moments of the departure process of the first queue. The problem in that paper is that the result of the moment matching did not define a valid stochastic process. To overcome this difficulty, [11] proposed to apply joint moment fitting instead of matching, relying on procedure [4]. Here, an alternative procedure based on Sect. 12.6.2.2 and the nonlinear program above is applied to solve the same problem.

The matrices of the lag-1 joint moments of the departure process corresponding to type-1 ($\boldsymbol{N}^{(1)}$) and type-2 ($\boldsymbol{N}^{(2)}$) jobs are

$$\begin{aligned} \boldsymbol{N}^{(1)} &= \begin{bmatrix} 0.7500 & 1.4959 & 11.2588 \\ 1.5001 & 4.6692 & 43.7821 \\ 11.0045 & 43.4938 & 438.3192 \end{bmatrix}, \\ \boldsymbol{N}^{(2)} &= \begin{bmatrix} 0.2500 & 0.5042 & 3.8188 \\ 0.5000 & 1.5330 & 14.2786 \\ 4.0731 & 14.8957 & 146.7858 \end{bmatrix}. \end{aligned} \tag{12.45}$$

Based on these joint moments, the moment matching method returns a MRAP given by matrices

$$H_0 = \begin{bmatrix} -1.4452 & 1.7636 & -2.6186 \\ 0.0160 & -0.9806 & 0.7218 \\ -0.3493 & 0.6472 & -1.0551 \end{bmatrix}, H_1 = \begin{bmatrix} 0.7311 & 0.6049 & 0.2228 \\ 0.0800 & 0.0795 & 0.0872 \\ 0.1708 & 0.1694 & 0.1630 \end{bmatrix},$$
$$H_2 = \begin{bmatrix} 0.3164 & 0.2776 & 0.1475 \\ -0.0044 & -0.0030 & 0.0034 \\ 0.0869 & 0.0873 & 0.0797 \end{bmatrix},$$

which is clearly a non-Markovian representation. Our procedure obtained an approximate MMAP based on the squared distance optimization, with matrices

$$D_0 = \begin{bmatrix} -0.18727 & 0 & 0 \\ 0 & -0.7735 & 0 \\ 0 & 0 & -2.5183 \end{bmatrix}, D_1 = \begin{bmatrix} 0.089235 & 0.027462 & 0.014315 \\ 0.034411 & 0.4187 & 0.18986 \\ 0.27113 & 0.78772 & 0.52658 \end{bmatrix},$$
$$D_2 = \begin{bmatrix} 0.045767 & 1.455 \times 10^{-5} & 0.010476 \\ 0.012185 & 0.066362 & 0.051978 \\ 0.060869 & 0.65762 & 0.21443 \end{bmatrix}.$$

The joint moments of the approximate MMAP are

$$\hat{N}^{(1)} = \begin{bmatrix} 0.74692 & 1.4461 & 10.609 \\ 1.4771 & 4.2299 & 38.164 \\ 10.763 & 38.644 & 376.84 \end{bmatrix}, \hat{N}^{(2)} = \begin{bmatrix} 0.25308 & 0.55401 & 4.4688 \\ 0.523 & 1.8518 & 18.427 \\ 4.315 & 18.093 & 188.1 \end{bmatrix},$$

which are relatively close to the original ones given by (12.45).

12.7 Application: Approximating the Departure Process of a MAP/MAP/1 Queue by a MAP

A popular approach for the analysis of a network of MAP/MAP/1 queues is the so called traffic-based decomposition, where the internal traffic in the network is modelled by MAPs. The closeness properties of MAPs over splitting and superposition make them ideal for this purpose. The key question is how to obtain a MAP that represents the departure process of a queue. Two options from the past literature which are known to perform relatively well are as follows:

- The ETAQA truncation of the queue length process in [23],
- and the joint moments-based procedure presented in [9].

In the practice both methods can return a RAP instead of a MAP, thus the procedure described in Sect. 12.6 becomes relevant.

12.7.1 Introduction to the Departure Process Analysis

The MAP/MAP/1 queue is a subclass of QBD queues, which are characterized by four matrices, $\boldsymbol{B}$, $\boldsymbol{F}$, $\boldsymbol{L}$, and $\boldsymbol{L_0}$. Matrices $\boldsymbol{B}$ and $\boldsymbol{F}$ consist of phase transition rates accompanied by service and arrival events, respectively, while matrices $\boldsymbol{L_0}$ and $\boldsymbol{L}$ correspond to the internal transitions when the queue is at level 0 and at level above zero. The generator matrix of the CTMC keeping track of the number of jobs in the queue and the phase of the system has a tri-diagonal structure given by

$$Q = \begin{bmatrix} \boldsymbol{L_0} & \boldsymbol{F} & & & \\ \boldsymbol{B} & \boldsymbol{L} & \boldsymbol{F} & & \\ & \boldsymbol{B} & \boldsymbol{L} & \boldsymbol{F} & \\ & & \ddots & \ddots & \ddots \end{bmatrix}. \tag{12.46}$$

Separating the transitions that generate a departure leads to a MAP that captures the departure process in an exact way as

$$\boldsymbol{D_0} = \begin{bmatrix} \boldsymbol{L_0} & \boldsymbol{F} & & & \\ & \boldsymbol{L} & \boldsymbol{F} & & \\ & & \boldsymbol{L} & \boldsymbol{F} & \\ & & & \ddots & \ddots \end{bmatrix}, \boldsymbol{D_1} = \begin{bmatrix} & & & \\ \boldsymbol{B} & & & \\ & \boldsymbol{B} & & \\ & & \ddots & \end{bmatrix}, \tag{12.47}$$

but unfortunately this representation has infinitely many states. A finite representation can be obtained by truncating the infinite model. It is proven in [23] that an appropriate truncation at level k is able to preserve the joint distribution of the departure process up to lag-$(k-1)$. The truncation at level k is done as

$$\boldsymbol{D}_0^{(k)} = \left[\begin{array}{cccc|c} \boldsymbol{L_0} & \boldsymbol{F} & & & 0 \\ & \boldsymbol{L} & \boldsymbol{F} & & 1 \\ & & \ddots & \ddots & \vdots \\ & & & \boldsymbol{L}+\boldsymbol{F} & k \\ & & & & \end{array}\right], \boldsymbol{D}_1^{(k)} = \left[\begin{array}{cccc|c} & & & & 0 \\ \boldsymbol{B} & & & & 1 \\ & \ddots & & & \vdots \\ & & \boldsymbol{B}-\boldsymbol{FG} & \boldsymbol{FG} & k \\ & & & & \end{array}\right], \tag{12.48}$$

where matrix $\boldsymbol{G}$ is the minimal nonnegative solution to the matrix-quadratic equation $\boldsymbol{0} = \boldsymbol{B} + \boldsymbol{L}\boldsymbol{G} + \boldsymbol{F}\boldsymbol{G}^2$.

Although the truncation leads to a finite model, the number of states can still be too large. The superposition operations in the queueing network increase the number of states even more, and the limits of numerical tractability are easily hit. A possible solution for the state-space explosion is provided in [9], where a compact representation is constructed while maintaining the lag-1 joint moments of the large process.

12.7.2 Practical Problems and Possible Solutions

An issue with both the ETAQA departure model and the joint moment-based approach is that they do not always return a Markovian representation; it is not even guaranteed that the departure model is a valid stochastic process.

Applying the RAP approximation procedure presented in Sect. 12.6 makes it possible to overcome this problem. Based on $(\boldsymbol{D}_0^{(k)}, \boldsymbol{D}_1^{(k)})$ it always returns a valid Markovian representation $(\boldsymbol{H}_0, \boldsymbol{H}_1)$, and at the same time it is also able to compress the truncated departure process to a desired level.

There is, however, one issue which has to be taken account when applying the procedure of Sect. 12.6, namely that the number of marginal moments that can be used to obtain matrix $(\boldsymbol{H}_0$ is limited. We are going to show that the order of the PH distribution representing the inter-departure times is finite (denoted by N_D), determined by $2N_D - 1$ moments, and using more moments during the approximation leads to a dependent moment set (see [6]).

Theorem 12.6 *The order of the PH distribution representing the inter-departure times of a QBD queue with block size $N > 1$ is*

$$N_D = 2N. \tag{12.49}$$

Proof In [23] it is shown how an order $2N$ PH distribution is constructed that captures the inter-departure times in an exact way, thus $N_D \leq 2N$. Additionally, it is easy to find concrete matrices $\boldsymbol{B}, \boldsymbol{F}, \boldsymbol{L}$ and $\boldsymbol{L}_0$ such that the order of this PH distribution is exactly $2N$ (practically any random matrices are suitable, the order can be determined by the STAIRCASE algorithm of [5]). Consequently, we have that $N_D = 2N$. □

Surprisingly, in case of MAP/MAP/1 queues the order of the inter-departure times is lower.

Theorem 12.7 ([9], Theorem 2) *The order of the PH distribution representing the inter-departure times of a MAP/MAP/1 queue is*

$$N_D = N_A + N_S, \tag{12.50}$$

where N_A denotes the size of the MAP describing the arrival process and N_S the one of the service process, assuming that $N_A + N_S > 1$.

Thus, the proposed method for producing a MAP $(\boldsymbol{B}_0, \boldsymbol{B}_1)$ that approximates the departure process is as follows:

1. First, the ETAQA departure model is constructed up to the desired lag k, providing matrices $(\boldsymbol{D}_0^{(k)}, \boldsymbol{D}_1^{(k)})$. The stationary phase distribution at departure

instants needs to be determined as well, α_D is the unique solution to $\alpha_D(-\boldsymbol{D}_0^{(k)})^{-1}\boldsymbol{D}_0^{(k)}, \alpha_D\mathbb{1} = 1$.

2. The marginal moments of the inter-departure times are computed from α_D and $\boldsymbol{D}_0^{(k)}$. The more moments are taken into account, the larger the output of the approximation is. According to the above theorems, more than $2N_D - 1$ should not be used.
3. Matrix $\boldsymbol{B}_0$ is obtained by moment matching (see Sect. 12.6.2.2).
4. Matrix $\boldsymbol{B}_1$ is obtained such that either the squared distance of the joint density is minimized up to lag k, see Sect. 12.6.2.1.

12.7.3 Numerical Example

In this example,[2] we consider a simple tandem queueing network of two MAP/MAP/1 queues. The arrival process of the first station is given by matrices

$$\boldsymbol{D}_0 = \begin{bmatrix} -0.542 & 0.003 & 0 \\ 0.04 & -0.23 & 0.01 \\ 0 & 0.001 & -2.269 \end{bmatrix}, \boldsymbol{D}_1 = \begin{bmatrix} 0.021 & 0 & 0.518 \\ 0 & 0.17 & 0.01 \\ 0.004 & 0.005 & 2.259 \end{bmatrix}, \tag{12.51}$$

while the matrices characterizing the service process are

$$\boldsymbol{S}_0 = \begin{bmatrix} -10 & 0 \\ 0 & -2.22 \end{bmatrix}, \boldsymbol{S}_1 = \begin{bmatrix} 7.5 & 2.5 \\ 0.4 & 1.82 \end{bmatrix}. \tag{12.52}$$

With these parameters both the arrival and the service times are positively correlated ($\rho_1^{(A)} = 0.21$ and $\rho_1^{(S)} = 0.112$) and the utilization of the first queue is 0.624.

The service times of the second station are Erlang distributed with order 2 and intensity parameter 6 leading to utilization 0.685.

This queueing network is analyzed such a way, that the departure process is approximated by the ETAQA truncation and by the joint moments-based methods. Next, our RAP approximation procedure (Sect. 12.6) is applied to address the issues of the approximate departure processes, namely to obtain a Markovian approximation and in case of the ETAQA truncation method, to compress the large model to a compact one.

Table 12.1 depicts the mean queue length of the second station and the model size by various departure process approximations. The ETAQA truncation model has been applied with truncation levels 2 and 6, which has been compressed by our

[2]The implementation of the presented method and all the numerical examples can be downloaded from http://www.hit.bme.hu/~ghorvath/software.

Table 12.1 Results of the queueing network example

Model of the departure process	#states	E(queue len.)
Accurate result (simulation)	n/a	2.6592
ETAQA, lag-1 truncation	18	2.3379
Our method based on 3 moments and $\mathcal{D}_2\{\}$	2	2.4266
Our method based on 5 moments and $\mathcal{D}_2\{\}$	3	2.5722
ETAQA, lag-5 truncation	42	2.5405
Our method based on 3 moments and $\mathcal{D}_2\{\}$	2	2.4266
Our method based on 5 moments and $\mathcal{D}_2\{\}$	3	2.5722
Our method based on 3 moments and $\mathcal{D}_6\{\}$	2	2.4266
Our method based on 5 moments and $\mathcal{D}_6\{\}$	3	2.6805
Joint moments based, 2 states	2	2.3255
Our method based on 3 moments and $\mathcal{D}_2\{\}$	2	2.3255
Joint moments based, 3 states	3	2.755
Our method based on 3 moments and $\mathcal{D}_2\{\}$	2	2.4266
Our method based on 5 moments and $\mathcal{D}_2\{\}$	3	2.7489

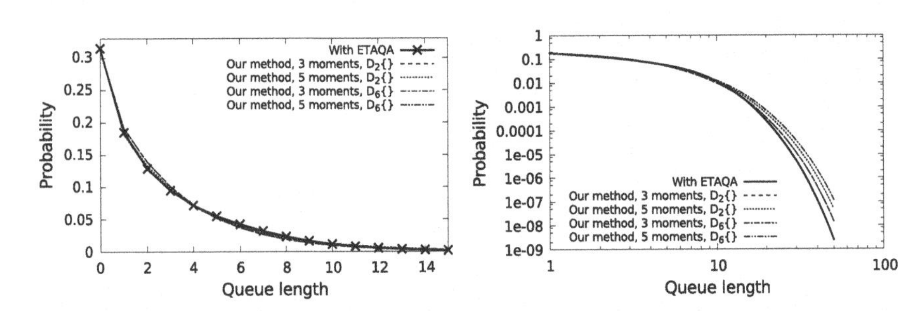

Fig. 12.8 Queue length distribution with the ETAQA departure model and its Markovian approximations

method based on either 3 or 5 marginal moments and with $\mathcal{D}_2\{\}$ or $\mathcal{D}_6\{\}$ distance optimization. The corresponding queue length distributions at the second station are compared in Fig. 12.8. The departure process has also been approximated by the joint moments-based method of [9], and an approximate Markovian representation has been constructed with our method based on 3 or 5 marginal moments and $\mathcal{D}_2\{\}$ optimization.

The results indicate that the RAP approximation and state-space compression technique presented in this chapter is efficient; the MAP returned is able to capture the important characteristic of the target RAP with an acceptable error.

Acknowledgment This work was supported by the Hungarian research project OTKA K101150 and by the János Bolyai Research Scholarship of the Hungarian Academy of Sciences.

References

1. Asmussen S, Bladt M (1999) Point processes with finite-dimensional conditional probabilities. Stochast Process Appl 82:127–142
2. Asmussen S, Koole G (1993) Marked point processes as limits of Markovian arrival streams. J Appl Probab 365–372
3. Bean NG, Nielsen BF (2010) Quasi-birth-and-death processes with rational arrival process components. Stochast Models 26(3):309–334
4. Buchholz Peter, Kemper Peter, Kriege Jan (2010) Multi-class Markovian arrival processes and their parameter fitting. Perform Eval 67(11):1092–1106
5. Buchholz Peter, Telek Miklós (2013) On minimal representations of rational arrival processes. Ann Oper Res 202(1):35–58
6. Casale G, Zhang EZ, Smirni E (2010) Trace data characterization and fitting for Markov modeling. Perform Eval 67(2):61–79
7. Golub GH, Nash S, Van Loan C (1979) A Hessenberg-Schur method for the problem AX + XB = C. IEEE Trans Autom Control 24(6):909–913
8. He Qi-Ming, Neuts Marcel (1998) Markov arrival processes with marked transitions. Stochast Process Appl 74:37–52
9. Horváth András, Horváth Gábor, Telek Miklós (2010) A joint moments based analysis of networks of MAP/MAP/1 queues. Perform Eval 67(9):759–778
10. Horváth G (2013) Moment matching-based distribution fitting with generalized hyper-Erlang distributions. In: Analytical and stochastic modeling techniques and applications, pp 232–246. Springer
11. Horváth Gábor, Van Houdt Benny (2013) Departure process analysis of the multi-type MMAP [K]/PH[K]/1 FCFS queue. Perform Eval 70(6):423–439
12. Horváth G (2015) Measuring the distance between maps and some applications. In: Gribaudo M, Manini D, Remke A (eds) Analytical and Stochastic Modelling Techniques and Applications, Lecture Notes in Computer Science, vol 9081, pp 100–114. Springer
13. Johnson MA, Taaffe MR (1989) Matching moments to phase distributions: mixtures of Erlang distributions of common order. Stochast Models 5(4):711–743
14. Latouche G, Ramaswami V (1987) Introduction to matrix analytic methods in stochastic modeling, volume 5. Soc Ind Appl Math
15. Laub AJ (2005) Matrix analysis for scientists and engineers. SIAM
16. Lipsky L (2008) Queueing theory: A linear algebraic approach. Springer
17. Neuts M (1975) Probability distributions of phase type. In: Liber Amicorum Prof. Emeritus H. Florin, pp 173–206. University of Louvain
18. Neuts MF (1979) A versatile Markovian point process. J Appl Probab 16:764–779
19. Rolski T, Schmidli H, Schmidt V, Teugels J (1999) Stochastic processes for finance and insurance. Willey, New York
20. Steeb WH (1997) Matrix calculus and Kronecker product with applications and C+ + programs. World Scientific
21. Telek Miklós, Horváth Gábor (2007) A minimal representation of Markov arrival processes and a moments matching method. Perform Eval 64(9):1153–1168
22. van de Liefvoort A (1990) The moment problem for continuous distributions. Technical report, University of Missouri, WP-CM-1990-02, Kansas City
23. Zhang Qi, Heindl Armin, Smirni Evgenia (2005) Characterizing the BMAP/MAP/1 departure process via the ETAQA truncation. Stochast Models 21(2–3):821–846

Chapter 13
Markovian Agent Models: A Dynamic Population of Interdependent Markovian Agents

Andrea Bobbio, Davide Cerotti, Marco Gribaudo, Mauro Iacono and Daniele Manini

Abstract A Markovian Agent Model (MAM) is an agent-based spatio-temporal analytical formalism aimed to model a collection of interacting entities, called Markovian Agents (MA), guided by stochastic behaviours. An MA is characterized by a finite number of states over which a transition kernel is defined. Transitions can either be local, or induced by the state of other agents in the system. Agents operate in a space that can be either continuous, or composed by a discrete number of locations. MAs may belong to different classes and each class can be parametrized depending on the location in the geographical (or abstract) space. In this work, we provide a very general analytical formulation of an MAM that encompasses many forms of physical dependencies among objects and many ways in which the spatial density may change in time. We revisit recent literature to show how previous works can be cast in terms of this more general MAM formulation.

Keywords Agent-based model · Spatially distributed systems · Performance modelling

13.1 Introduction

A Markovian Agent Model (MAM) is an agent-based spatio-temporal analytical formalism aimed to model a collection of interacting Markovian Agents (MAs). An MA has a finite number of possible discrete operating modes (states) over which

A. Bobbio
Università del Piemonte Orientale, Alessandria, Italy

D. Cerotti · M. Gribaudo (✉)
Politecnico di Milano, Milan, Italy
e-mail: marco.gribaudo@polimi.it

M. Iacono
Seconda Università di Napoli, Caserta, Italy

D. Manini
Università di Torino, Turin, Italy

K. Al-Begain and A. Bargiela (eds.), *Seminal Contributions to Modelling and Simulation*, Simulation Foundations, Methods and Applications,
DOI 10.1007/978-3-319-33786-9_13

a transition kernel is defined. The transition kernel is composed by two components a *local transition matrix* and an *induced transition matrix*. The local matrix contains a fixed component that depends on the MA structure and its position in the space but does not depend on the interaction with the other MAs. The induced matrix depends on the interaction of an MA with the other MAs. MAs may belong to different classes and each class has a spatial density function that depends on the location in the geographical (or abstract) space. In this work, we provide the solution equations for a very general MAM that encompasses a *static* MAM in which the spatial density of the MAs of any class remains constant in time and a *dynamic* MAM in which the spacial density varies as a function of the time. The density variation in the dynamic MAM can be due to the displacement of the MAs in the space (being constant their overall number) or to the birth or death of MAs. A typical example of a population of agents with similar characteristics is represented by the number of calls in a wireless cellular network where each cell can have a different number of ongoing calls, and this number may change due to new calls (birth), completion of calls (death) or and handoff-in and handoff-out calls from adjacent cells, or also due to reduction in the performance of a cell caused by degradation or failure [30].

The modelling and analysis of large-scale stochastic systems composed by spatially defined interacting objects has been mainly faced in the literature by resorting to the superposition of interacting Markov chains [1, 25], to stochastic Reward model or hierarchical models [10, 23], to various process algebra formalisms [15, 20] and to fluid models [18, 21].

In recent years, the MAM has emerged as a new versatile analytical technique whose main idea is to model a distributed system by means of interacting agents, so that each agent maintains its local properties but at the same time modifies its behaviour according to the influence of the interaction with the other agents [3, 17, 20]. By separating the local behaviour of an MA, that does not depend on the interaction with other agents, from the influence of the other MAs, we can avoid the construction of the combined state space. The solution of the overall model is obtained by building several models, one for each MA, and then solving them separately. This technique, that avoids generating a large model, is sometimes referred to as *largeness avoidance*. In the present work, we provide a very general analytical formulation of an MAM that encompasses many kinds of forms of physical dependencies among objects and many ways in which the spatial density may change in time. We revisit recent literature to show how previous works can be cast in terms of this more general MAM formulation.

Section 13.2 introduces the theory of MAM to arrive to the equations describing the dynamics of the probability distribution of a class of MAs as a function of the time and of the MA location in the space. Section 13.3 discusses the way in which the influence matrix is defined and evaluated for several different case studies. A wide spectrum of different cases is discussed in this section to give the reader a flavour of the possibilities offered by the technique.

13.2 Theory

Markovian Agents describe systems composed by several interacting agents, each one characterized by a set of possible *states*, and whose evolution can be described by the way in which they evolve in their state space. MAs behaviour is similar to Automata, and in particular to Continuous Time Markov Chains (CTMCs). The structure of a single MA is represented in Fig. 13.1. The circles labelled $i, j, \ldots k$, represent the states of the CTMC describing the MA. Differently from conventional CTMC models, the transitions among the states are of two possible types that are drawn differently:

- Solid lines (like the transition from i to j or the self-loops in i or in j) represent the local or autonomous behaviour of the objects. They are the fixed component of the infinitesimal generator that are independent of the interaction with the other MAs. For instance, they can be used to model a failure occurring to the agent, or the reaction to an internal stimulus. MAs include also self-loop transitions that require a particular notation since they are not visible in the infinitesimal generator of the CTMC [29]. Self-loop transitions are required because, even if they do not vary the state of the agent, they might influence the behaviour of other agents.
- Dashed lines (like the transition from i to k or the transitions into i or j) represent the transitions induced by the interaction with the other MAs. In the literature, several different types of induction have been considered. In any case, induction, as the name suggests, *induces* an agent to change state at a rate that depends on the state of the other agents. The way in which the rates of the induced transitions are computed is explained in Sect. 13.3.

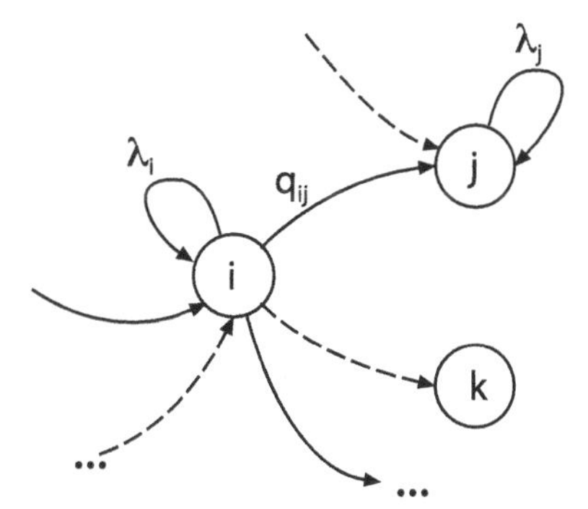

Fig. 13.1 Schematic structure of a Markovian agent

13.2.1 Markovian Agents Model

The Markovian Agent Model (MAM) represents a system as a collection of *Markovian Agents* (MAs) spread over a geographical space $\mathcal{V}$. As introduced, the essence of the MAM is that each MA is described by a CTMC, whose infinitesimal generator contains a fixed component that depends on the MA structure and position in space $\mathbf{v} \in \mathcal{V}$, and a component that depends on the interaction with the other MAMs.

Space $\mathcal{V}$ can be either discrete or continuous. If discrete, $\mathcal{V} = \{\mathbf{v_1}, \mathbf{v_2}, \ldots \mathbf{v_N}\}$ where $\mathbf{v_i}$ are the locations. If continuous, $\mathcal{V} \subseteq \mathbb{R}^k$, with $1 \leq k \leq 3$. Agents are divided in C classes, and an MA belonging to class c (with $1 \leq c \leq C$) is characterized by n_c different states. Let us call $p_j^{\{c\}}(t, \mathbf{v})$ the probability that a class c agent is in state $1 \leq j \leq n_c$ at time t, at location $\mathbf{v} \in \mathcal{V}$.

Agents are analyzed using *counting process* and exploiting mean field approximation [2, 24]. In particular, any location $\mathbf{v}$ can hold a number of class c agents that is defined by function $\rho_c(t, \mathbf{v})$. For discrete space models, $\rho_c(t, \mathbf{v})$ accounts for the average number of class c agents in location $\mathbf{v}$ at time t. For continuous space models, $\rho_c(t, \mathbf{v})$ is the average density of class c agents in point $\mathbf{v}$ at time t, and for any finite area $\Omega \subseteq \mathcal{V}$, the average number $N_c(t, \Omega)$ of class c agents contained inside Ω at time t can be computed as

$$N_c(t, \Omega) = \int_{\Omega} \rho_c(t, \mathbf{v}) d\mathbf{v}.$$

We also define $\pi_j^{\{c\}}(t, \mathbf{v}) = p_j^{\{c\}}(t, \mathbf{v}) \cdot \rho_c(t, \mathbf{v})$ as the density of class c agents in state j at location $\mathbf{v}$ and time t. Note that if we are considering a discrete model, where each location has exactly one agent, we have $\pi_j^{\{c\}}(t, \mathbf{v}) = p_j^{\{c\}}(t, \mathbf{v})$. Value $\pi_j^{\{c\}}(t, \mathbf{v})$ will be the main performance index that will be exploited to compute all the interesting measures in the model solution process.

According to the definition of the density $\rho_c(t, \mathbf{v})$, we can then classify MAMs with the following taxonomy:

- An MAM is *static* if $\rho(t, \mathbf{v})$ does not depend on time, and *dynamic* otherwise;
- An MAM is *discrete* if the geographical area on which the MAs are deployed is discretized and $\rho(t, \mathbf{v})$ is a discrete function of the space or it is *continuous* if $\rho(t, \mathbf{v})$ is a continuous function of the space;
- An MAM is *single class* if $C = 1$, and *multi class* if $C > 1$.

To simplify the notation, let us collect the terms into row vectors $\pi_c(t, \mathbf{v}) = |\pi_j^{\{c\}}(t, \mathbf{v})|$ that represents the state distribution of an MA belonging to class c at time t in position $\mathbf{v}$. Moreover, let $\Pi_{\mathcal{V}}(t) = \{(c, \mathbf{v}, \pi_c(t, \mathbf{v})) : 1 \leq c \leq C, \mathbf{v} \in \mathcal{V}\}$ be the ensemble of the probability distribution of all the agents of all the classes at time

t, and let us denote with $[\Pi_{\mathcal{V}}] = \{(t', \Pi_{\mathcal{V}}(t')) : 0 \leq t' < t\}$ the evolution of the ensemble of the probability distribution from the initial state up to time t. The dynamic of the state density distribution of a class c agent in position $\mathbf{v}$ is described by the following equation:

$$\frac{\partial \pi_c(t, \mathbf{v})}{\partial t} + \frac{\partial(\omega_c(t, \mathbf{v}, [\Pi_{\mathcal{V}}]) \cdot \pi_c(t, \mathbf{v})^T)}{\partial \mathbf{v}} = \nu_c(t, \mathbf{v}, [\Pi_{\mathcal{V}}]) + \pi_c(t, \mathbf{v}) \cdot \mathbf{K}_c(t, \mathbf{v}, [\Pi_{\mathcal{V}}]). \tag{13.1}$$

The terms $\omega_c(t, \mathbf{v}, [\Pi_{\mathcal{V}}])$, $\nu(t, \mathbf{v}, [\Pi_{\mathcal{V}}])$ and $\mathbf{K}_c(t, \mathbf{v}, [\Pi_{\mathcal{V}}])$ are respectively the *motion*, *increase* and *transition* kernels. They can all depend on the class c, on the position $\mathbf{v}$, on the time t and on the ensemble probability $[\Pi_{\mathcal{V}}]$, and they rule the evolution of the system.

The motion kernel $\omega_c(t, \mathbf{v}, [\Pi_{\mathcal{V}}])$ is used only for agents deployed in continuous space: for models that are characterized by a finite number of locations, the second term on l.h.s. of Eq. (13.1) can be discarded since it is not required. Otherwise $\omega_c(t, \mathbf{v}, [\Pi_{\mathcal{V}}]) = (\omega_x^{\{c\}}(t, \mathbf{v}, [\Pi_{\mathcal{V}}]), \omega_y^{\{c\}}(t, \mathbf{v}, [\Pi_{\mathcal{V}}]), \ldots)$ is a set of diagonal matrices whose element $\omega_x^{\{c:i\}}(t, \mathbf{v}, [\Pi_{\mathcal{V}}])$ represents the speed along direction x for a class c agent in state i (with $1 \leq i \leq n_c$). With a slight abuse of notation, $\frac{\partial}{\partial v} = \left(\frac{\partial}{\partial x}, \frac{\partial}{\partial y}, \ldots\right)$ denotes the partial derivates along all the dimensions of space V. For example, for a two-dimensional system, we have:

$$\begin{aligned}\frac{\partial(\omega_c(t, \mathbf{v}, [\Pi_{\mathcal{V}}]) \cdot \pi_c(t, \mathbf{v})^T)}{\partial v} = & \frac{\partial(\omega_x^{\{c\}}(t, \mathbf{v}, [\Pi_{\mathcal{V}}]) \cdot \pi_c(t, \mathbf{v})^T)}{\partial x} \\ & + \frac{\partial(\omega_y^{\{c\}}(t, \mathbf{v}, [\Pi_{\mathcal{V}}]) \cdot \pi_c(t, \mathbf{v})^T)}{\partial y}.\end{aligned} \tag{13.2}$$

The increased kernel $\nu_c(t, \mathbf{v}, [\Pi_{\mathcal{V}}])$ accounts for the effects that grows the number or the density of agents in a point in space. It can be subdivided into two terms:

$$\nu_c(t, \mathbf{v}, [\Pi_{\mathcal{V}}]) = b_c(t, \mathbf{v}, [\Pi_{\mathcal{V}}]) + m_c^{in}(t, \mathbf{v}, [\Pi_{\mathcal{V}}]). \tag{13.3}$$

Term $b_c(t, \mathbf{v}, [\Pi_{\mathcal{V}}])$ is the *birth* term: it is used to model the autonomous generation of agents and expresses the rate (measured in agents per time unit or agent density per time unit) at which class c agents are created in location $\mathbf{v}$ at time t. Term $\mathbf{m}_c^{in}(t, \mathbf{v}, [\Pi_{\mathcal{V}}])$ is the location *input* term, and accounts for class c agents that "warp" into location $\mathbf{v}$ at time t from other points in space. In discrete space model, it is used to model cell transfer of agents. Continuous space models may use this term to account for "wormhole" or "portal" effects where agents that reach one location, can vanish and immediately reappear in other points in the space.

The transition kernel $\mathbf{K}_c(t, \mathbf{v}, [\Pi_\mathcal{V}])$ accounts for both the state transitions of the agents, and for the effects that can reduce the number of agents in one location. It can be subdivided into four terms:

$$\begin{aligned}\mathbf{K}_c(t, \mathbf{v}, [\Pi_\mathcal{V}]) = \mathbf{Q}_c(t, \mathbf{v}) + I_c(t, \mathbf{v}, [\Pi_\mathcal{V}]) + \\ - \mathbf{D}_c(t, \mathbf{v}, [\Pi_\mathcal{V}]) - \mathbf{M}_c^{out}(t, \mathbf{v}, [\Pi_\mathcal{V}]).\end{aligned} \tag{13.4}$$

Matrix $\mathbf{Q}_c(t, \mathbf{v}) = |q_{ij}^{\{c\}}(t, \mathbf{v})|$ in Eq. (13.1) defines the rate of *local transitions*, as in traditional CTMC, according to the agent position $\mathbf{v}$ and time t. Note that to further emphasize the locality of this term, $\mathbf{Q}_c(t, \mathbf{v})$ does not depend on the complete state history of the model $[\Pi_\mathcal{V}]$. The influence matrix $\mathcal{I}_c(t, \mathbf{v}, [\Pi_\mathcal{V}])$ accounts for the rate of *induced transitions* due to the influence of other agents. The entries of matrix $\mathcal{I}_c(t, \mathbf{v}, [\Pi_\mathcal{V}])$ depend on the state probabilities of other agents and must satisfy precise structural restrictions so that the matrix $\mathbf{Q}_c(t, \mathbf{v}) + \mathcal{I}_c(t, \mathbf{v}, [\Pi_\mathcal{V}])$ is still an infinitesimal generator matrix. Matrix $\mathbf{D}_c(t, \mathbf{v})$ is the *death* term: it is a diagonal matrix whose elements $d_{ii}^{\{c\}}(t, \mathbf{v})$ represent the rate at which class c agent in location $\mathbf{v}$ at time t may disappear from the model depending on its state i. Matrix $\mathbf{M}_c^{out}(t, \mathbf{v}, [\Pi_\mathcal{V}])$ is a matrix whose terms $m_{ij}^{out:\{c\}}(t, \mathbf{v})$ account for the discrete output from a location $\mathbf{v}$ at time t for a class c agent. **If** $i = j$, output does not cause a change of state; otherwise, the state of the agent may vary during its motion. It is the exit counterpart of vector $\mathbf{m}_c^{in}(t, \mathbf{v}, [\Pi_\mathcal{V}])$ previously introduced. In particular, the two terms are usually related by some conservation law. For example, if agents can warp from location $\mathbf{v}$ to location $\mathbf{u}$ at rate λ, we have $\mathbf{m}_c^{in}(t, \mathbf{u}, [\Pi_\mathcal{V}]) = |\lambda, \ldots|\pi_c(t, \mathbf{v})$ and $\mathbf{M}_c^{out}(t, \mathbf{v}, [\Pi_\mathcal{V}]) = diag(\lambda, \ldots)$.

In order to solve the model, the initial state of the system must be provided. For discrete space models, it is enough to provide $\rho_c(0, \mathbf{v})$, the initial density of class c agents in location $\mathbf{v}$, and $p_j^{\{c\}}(0, \mathbf{v})$, the corresponding initial state probability. From $p_j^{\{c\}}(0, \mathbf{v})$ and $\rho_c(0, \mathbf{v})$ we can then compute the initial condition of Eq. (13.1), and express it as:

$$\pi_j^{\{c\}}(\mathbf{v}, \nu) = p_j^{\{c\}}(0, \mathbf{v}) \cdot \rho_c(0, \mathbf{v}). \tag{13.5}$$

Continuous space models also require boundary conditions to describe what happens to agents that exit the domain V, and whether external agents enters inside V from outside. To simplify the presentation, we will limit our discussion to cases in which agents are confined inside V. In this case the boundary conditions are:

$$\rho_c(t, \mathbf{v}) = 0, \forall \mathbf{v} \in \mathbf{Boundary}(\mathcal{V}), \tag{13.6}$$

where $\mathbf{Boundary}(\mathcal{V})$ is the set of all points that define the boundary of the considered space $\mathcal{V}$.

13.3 Induced Transitions

Equation (13.1) provides an abstract description of the agents' behaviour, since the rates of induced transitions composing the influence matrix are not fully specified. The definition and evaluation of the influence matrix depend on the considered problem. In this section, we will present several cases from the literature, and we will show how they can be incorporated in Eq. (13.1).

13.3.1 *Largeness Avoidance: Shared Repair in Complex Systems*

Large fault-tolerant systems are composed by many subsystems that may undergo failures and repairs during operation. High availability is a prescribed requirement in many applications and necessitates an accurate model. To capture dynamic behaviour and dependencies that arise in the failure and repair process of the system, a preferred way is to resort to state-space models. However, state-space models suffer from state-space explosion; i.e. extremely large state-space is required for the accurate modelling of real systems, referred to as *largeness*. A way to tackle largeness is to avoid generating a large model, called *largeness avoidance*. Equation 13.1 offers a technique of *largeness avoidance*. Each subsystem is modelled as an MA with its local infinitesimal generator $\mathbf{Q}_c(t)$ (there is no spatial dependence on the position $\mathbf{v}$, in this case) and the influence matrix $\mathcal{I}_c(t, [\Pi_{\mathcal{V}}])$ that must be inferred from the physical dependencies inside the system. The *largeness avoidance* is obtained by computing the overall system dependability measures by solving individual MAs defined on the state-space of each subsystem. Shared repair is an usual source of dependence, and examples of availability modelling with various kinds of dependencies due to shared repair are given in [27, 28].

A simple example, inspired from [27], is the following. A system is composed by n subsystems that for simplicity we represent as a two-state model with a single up state u and a single down state d. The subsystems share a single repair person that follows a *preemptive repair priority policy*; i.e. the subsystems are ordered according to a predefined repair list, that for the sake of simplicity we assume in the natural order $X_1 \succ X_2 \succ \cdots \succ X_n$, and upon failure the repair crew starts repairing the subsystems with higher priority (subsystem X_i before X_j with $j > i$). Let us call λ_i the failure rate of component i, and μ_i the corresponding repair rate. We consider here only the steady-state solution.

Given that X_1 is repaired first, the repair of the other subsystems with lower priority is delayed. We can account for this delay modifying the repair rates of the subsystems with lower priority. The MA related to model X_1 is solved first and the probability $\pi_d^{X_1}$ that the subsystem X_1 is under repair (i.e. is in the down state) is calculated. Thus the probability that the repair person is idle is $\pi_u^{X_1} = (1 - \pi_d^{X_1})$. Subsystem X_2 can be repaired only if the repair person is not busy with subsystem

X_1 (i.e. X_1 is in the up state), which can be accounted for by reducing the repair rate of subsystem X_2 by the factor $\pi_u^{X_1}$:

$$\mu_2' = \mu_2 \pi_u^{X_1} = \mu_2 (1 - \pi_d^{X_1}).$$

By the same reasoning, subsystem X_j can be repaired only if all the subsystems $X_i, (i<j)$ are not under repair and are operating. This effect is accounted for by reducing the repair rate of subsystem X_j by:

$$\mu_j' = \mu_j \prod_{i<j} \pi_u^{X_i} = \mu_j \prod_{i<j} (1 - \pi_d^{X_i}).$$

The above problem can be rewritten in terms of MAM using Eq. (13.4). Each subsystem is an MA whose local kernel contains only the failure transitions that are independent of the rest of the system:

$$\mathbf{Q}^{X_j} = \begin{bmatrix} -\lambda_j & \lambda_j \\ 0 & 0 \end{bmatrix} \quad \text{for} \quad \text{any} \quad j.$$

The influence matrix accounts for the repair transitions that are influenced by the other agents:

$$\mathcal{I}^{X_j} = \begin{bmatrix} 0 & 0 \\ \mu_j \prod_{i<j} \pi_u^{X_i} & -\mu_j \prod_{i<j} \pi_u^{X_i} \end{bmatrix}.$$

And for each MA:

$$\mathrm{K}^{X_j} = \mathbf{Q}^{X_j} + \mathcal{I}^{X_j}.$$

In this very simple case, the influence matrix can be computed sequentially from $\mathcal{I}^{X_j}$ to $\mathcal{I}^{X_{j+1}}$. In more complex repair schemes [27, 28], the computation of the influence matrix requires a fixed-point iteration.

13.3.2 Message Passing Model

In the message passing model, the influence among MAs is represented by the exchange of relational entities, called *messages*, that are emitted by an MA and perceived by the other ones modifying their stochastic dynamics. The interaction among agents is ruled by a *perception function* that captures the sending and receiving aptitude of the involved MAs and is a function of their geographical location and of the features of the traversed media. MAs may belong to different classes with different local behaviours and interaction capabilities, and messages may belong to different types where each type induces a different effect on the interaction mechanism. The perception function describes how a message of a given type emitted by an MA of a given class in a given position in the space is perceived by an MA of a given class in a different position.

In particular, the message passing model considers a set of different messages $\mathcal{M}$, and a set $\mathcal{U} = \{u_1(\cdot)\ldots u_{|\mathcal{M}|}(\cdot)\}$ of $|\mathcal{M}|$ perception functions, one for each message type.

This model does not allow birth or death of agents ($\mathrm{b}_c(t, \mathbf{v}, [\Pi_\mathcal{V}]) = 0$ and $\mathbf{D}_c(t, \mathbf{v}, [\Pi_\mathcal{V}]) = 0$). Matrix $\mathbf{Q}_c(\mathbf{v})$ is the $n_c \times n_c$ infinitesimal generator matrix of the CTMC that describes the local behaviour of a class c agent in position $\mathbf{v}$, and corresponds to the local transition kernel of the general model. In order to consider the possibility to send messages without changing state, self-jump transitions are explicitly included in the model. In particular, the n_c component vector $\mathbf{\Lambda}_c(\mathbf{v}) = |\lambda_j^{\{c\}}(\mathbf{v})|$ represents the rates of *self-jumps* for a class c agent in position $\mathbf{v}$, i.e. the rates at which the CTMC reenters the same state.

The induced transition kernel $\mathcal{I}_c(t, \mathbf{v}, [\Pi_\mathcal{V}])$ is instead built starting from several parameters that characterize the way in which messages are sent and received. In particular, $\mathbf{G}_c(\mathbf{v}, \mu) = |g_{ij}^{\{c\}}(\mathbf{v}, \mu)|$ is a $n_c \times n_c$ matrix describing the probability that an agent of class c in position $\mathbf{v}$ generates a message of type $\mu \in 1\ldots|\mathcal{M}|$ during a jump from state i to state j, and $\mathbf{A}_c(\mathbf{v}, \mu) = |a_{ij}^{\{c\}}(\mathbf{v}, \mu)|$ is a $n_c \times n_c$ matrix, that describes the action activated upon acceptance of a type μ message for an agent of class c in position $\mathbf{v}$.

The perception function $u_\mu(\mathbf{v}, c, i, \mathbf{v}', c', i') \in [0, +\infty)$ represents the aptitude with which an agent of class c, in position $\mathbf{v}$, and in state i, perceives a message of type μ generated by an agent of class c' in position v' in state i'.

From the previous terms, we can define $\beta_j^{\{c\}}(\mathbf{v}, \mu)$ as the total rate at which messages of type μ are generated by an agent of class c in state j and in position $\mathbf{v}$:

$$\beta_j^{\{c\}}(\mathbf{v}, \mu) = \underbrace{\lambda_j^{\{c\}}(\mathbf{v}) g_{jj}^{\{c\}}(\mathbf{v}, \mu)}_{ⓐ} + \underbrace{\sum_{k \neq j} q_{jk}^{\{c\}}(\mathbf{v}) g_{jk}^{\{c\}}(\mathbf{v}, \mu)}_{ⓑ}, \tag{13.7}$$

where the first term (a) in the r.h.s is the contribution of the messages of type μ emitted during a self-loop from state j and the second term (b) is the contribution of messages of type μ emitted during a transition from state j to any state $k(\neq j)$.

Then we define $\gamma_{ii}^{\{c\}}(t, \mathbf{v}, [\Pi_\mathcal{V}], \mu)$ as the total rate at which messages of type μ coming from the whole volume V are perceived by an agent of class c, in state i, in position $\mathbf{v}$, at time t:

$$\gamma_{ii}^{\{c\}}(t, \mathbf{v}, [\Pi_\mathcal{V}], \mu) = \int_{\substack{\mathbf{v}' \in \mathcal{V} \\ \mathbf{v}' \neq \mathbf{v}}} \sum_{c'=1}^{C} \sum_{j=1}^{n_{c'}} u_\mu(\mathbf{v}, c, i, \mathbf{v}', c', j) \beta_j^{\{c'\}}(\mathbf{v}', \mu) \pi_j^{\{c'\}}(t, \mathbf{v}') d\mathbf{v}'. \tag{13.8}$$

The term $u_\mu(\mathbf{v}, c, i, \mathbf{v}', c', j)\beta_j^{\{c'\}}(\mathbf{v}', \mu)\pi_j^{\{c'\}}(t, \mathbf{v}')$ in Eq. (13.8) is the rate of type μ messages received by class c agent in state i in position $\mathbf{v}$, coming from a class c' agent in position $\mathbf{v}'$ in state j, at time t. Since this term depends on the current density of agents in other locations of the model $\pi_j^{\{c'\}}(t, \mathbf{v}')$, the value of $\gamma_{ii}^c(t, \mathbf{v}, [\Pi_\mathcal{V}], \mu)$ depends also on the total state of the model $[\Pi_\mathcal{V}]$. The total rate $\gamma_{ii}^c(t, \mathbf{v}, [\Pi_\mathcal{V}], \mu)$ is obtained integrating the contributions coming from all the states and all the agent classes, and over the entire area $\mathcal{V}$. Note that in discrete space models, the integral is replaced by a summation. We collect the rates of Eq. (13.8) in a diagonal matrix $\mathbf{\Gamma}_c(t, \mathbf{v}, [\Pi_\mathcal{V}], \mu) = \text{diag}(\gamma_{ii}^{\{c\}}(t, \mathbf{v}, [\Pi_\mathcal{V}], \mu))$. This matrix can be used to compute the induced transitions kernel $\mathcal{I}_c(t, \mathbf{v}, [\Pi_\mathcal{V}])$:

$$\mathcal{I}_c(t, \mathbf{v}, [\Pi_\mathcal{V}]) = \sum_{\mu \in \mathcal{M}} \mathbf{\Gamma}^{\{c\}}(t, \mathbf{v}, [\Pi_\mathcal{V}], \mu)\left(\mathbf{A}^{\{c\}}(\mathbf{v}, \mu) - \mathbf{I}\right). \tag{13.9}$$

In Eq. (13.9) the acceptance matrix $\mathbf{A}^{\{c\}}(\mathbf{v}, \mu)$ is used to decide whether message μ received by an agent in state i will cause a transition (with probability $1 - a_{ii}^{\{c\}}(\mathbf{v}, \mu)$), and to which state j its reception will lead ($a_{ij}^{\{c\}}(\mathbf{v}, \mu)$).

The message passing model has been applied in the literature to several case studies. For example, in the field of Wireless Sensor Networks, it has been used in [17] to study on-off policies, and in [3] to evaluate swarm intelligence based routing algorithms. In [6] messages are used to study the propagation of earthquakes, while in [9] they are used to study the propagation of fire. Finally, in [8] agent are used at two levels: to model both a physical phenomenon (fire propagation in forest) and a WSN monitoring infrastructure.

13.3.3 Spatial Density Dependent Communications

In the message passing model presented in Sect. 13.3.2, the value of the perception function depends on the type of message exchanged between two MAs, a sender and a receiver, and on their properties (location, class, and current state), as formally described by the notation $u_\mu(\mathbf{v}, c, i, \mathbf{v}', c', i')$. However, this dependency causes a one-to-one relationship that restricts the expressiveness of the model since the influence matrix $\mathcal{I}_c(t, \mathbf{v}[\Pi_\mathcal{V}])$ would in principle allow to represent one-to-all interactions. An effective trade-off between these extremes is given by considering one-to-many relations, where several neighbours of the perceiver agent are taken into account. In the following, we will describe an example of such approach to model the *Double Bridge Experiment* [12, 16], a famous ant colony optimization (ACO) problem [13].

In the experiment, two bridges connect a nest of ants with a food source. Two scenarios are investigated: in the former the lengths of the bridges are equal, in the latter a bridge is shorter than the other one. The experiment shows that using *stigmergy*, a form of indirect communication through the environment, the ant colony is able to reach the food following the shortest path. Indeed, during the journey from the nest to the food and vice versa, ants release an amount of a chemical substance called *pheromone*. Ants perceive it and are conditioned to choose with greater probability a path marked with a strong concentration of pheromone. In presence of bridges of different length, the ants choosing the shortest one reach the food earlier than those choosing the longer path. Thus, the pheromone trail increases faster on the shorter bridge and further ants choose it to reach food. In the end, the ant colony converges to follow the shortest path.

The dynamic of the ant colony in this experiment can be described by a simple stochastic model [12, 16], where the probability of choosing a given branch is given by:

$$p_{is}(t) = \frac{(k+\varphi_{is}(t))^{\alpha}}{(k+\varphi_{is}(t))^{\alpha}+(k+\varphi_{il}(t))^{\alpha}}, \quad p_{il}(t) = 1 - p_{is}(t), \tag{13.10}$$

where $p_{is}(t)$ is the probability of choosing the shorter branch, and $\varphi_{is}(t)$ is the amount of pheromone on the shorter branch at a time t. The same values on the longer branch are given by $p_{il}(t)$ and $\varphi_{il}(t)$. The parameter k is needed to provide a non-null probability of choosing a path not yet marked by pheromone; the exponent α provides a non-linear behaviour. We can observe that in such case, the probability of choosing a specific direction depends on the spatial distribution of the pheromone concentrations.

The MAM of the *Double Bridge Experiment* proposed in [4] represents ants as messages, and locations that ants traverse by MAs. More formally, agents are deployed on a geographical space $\mathcal{V}$ structured as an undirected graph $G = (\mathcal{V}, E)$ with $\mathcal{V} = \{\mathbf{v}_1, \mathbf{v}_2, \ldots \mathbf{v}_N\}$ the set of locations where the MAs reside, and E the edges of the graph. Figure 13.2 shows the graphs for the two scenarios of the experiments, where nodes labelled n and f represent the location of nest and source food, respectively. Messages passing from a location to another are depicted as little arrows with a label indicating the direction of the ants. Label m_{fw} for ants directed to the source food, m_{bw} for ants coming back to the nest. The number of hops from node n to node f represents the length of the path, thus Fig. 13.2a depicts a scenario with equal branches, whereas Fig. 13.2b the different branches one.

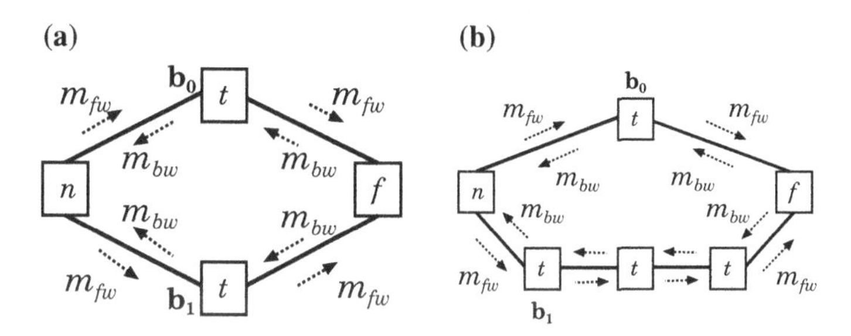

Fig. 13.2 Graph used to model the experiment scenarios. **a** Equal branches, **b** two different branches

The amount of pheromone is discretized in a finite number of levels and the MA in location $\mathbf{v}$ codes in its state space the current level of pheromone in that location. The arrival of ants causes the increment of the pheromone and is represented in the model by the reception of messages and consequent transition from a lower to an higher level state of the MA in location $\mathbf{v}$. In the model the birth and death of MAs are not considered (i.e. $\mathbf{b}_c(t, \mathbf{v}, [\Pi_{\mathcal{V}}]) = 0$ and $\mathbf{D}_c(t, \mathbf{v}, [\Pi_{\mathcal{V}}]) = 0$). Moreover, since the flow of messages represent the moving ants across a set of fixed locations, the term $\omega(t, \mathbf{v}, [\Pi_{\mathcal{V}}])$ is neglected too. Thus, the general Eq. (13.1) became

$$\frac{\partial \pi(t, \mathbf{v})}{\partial t} = \pi(t, \mathbf{v}) \cdot \mathbf{K}(t, \mathbf{v}, \pi_{\mathcal{V}}(t)), \tag{13.11}$$

where the transition kernel $\mathbf{K}(t, \mathbf{v}, \pi_{\mathcal{V}}(t))$ depends on the ensemble of the probability distribution of all agents at current time t only. In particular, it can be computed as described in Sect. 13.3.2 where the perception function $u_\mu(\cdot)$ with $\mu \in \{m_{fw}, m_{bw}\}$ is defined as

$$u_\mu(\mathbf{v}, \mathbf{v}', t) = \frac{(k + E[\pi(t, \mathbf{v})])^\alpha}{\sum_{\mathbf{v}'' \in Next^\mu(\mathbf{v}')} (k + E[\pi(t, \mathbf{v}'')])^\alpha}. \tag{13.12}$$

Parameters k and α are the same of Eq. (13.10), $E[\pi(t, \mathbf{v})]$ gives the mean value of the concentration of pheromone at a time t in position $\mathbf{v}$. The function $Next(\mathbf{v}')$ gives the set of elements $\{\mathbf{v}''\}$ such that the agent in position $\mathbf{v}''$ perceives a message emitted by the agent of in position $\mathbf{v}'$. Note that, even if the perception function is defined on a pair of sender and receiver MAs as in Sect. 13.3.2, its value depends on the properties of a set of agents that in this case are the neighbours of the sender MA.

13.3.4 Agent Motion Models

In Agent Motion Models (AMMs), the MAs interact each others by exchanging messages, and move across a continuous geographical space $\mathcal{V} \subseteq \mathbb{R}^k$ with $k \leq 3$. For simplicity in the following we apply some restrictions, in particular: (i) we focus on single-class AMMs, thus dropping the subscript c and superscript $\{c\}$ in the equations, (ii) we consider a one-dimensional geographical space, i.e. a straight line, by setting $k = 1$ and (iii) we exclude "warp" phenomena by setting both terms $\mathbf{m}^{in}(t, \mathbf{v}, [\pi_{\mathcal{V}}]) = 0$ and $\mathbf{M}^{out}(t, \mathbf{v}, [\pi_{\mathcal{V}}]) = 0$.

With such assumptions, Eq. (13.1) became

$$\frac{\partial \pi(t, x)}{\partial t} + \frac{\partial(\omega(t, x, [\pi_{\mathcal{V}}]) \cdot \pi_c(t, x)^T)}{\partial x} = \mathbf{b}(t, v, [\pi_{\mathcal{V}}]) + \pi(t, x) \cdot \mathbf{K}(t, x, [\pi_{\mathcal{V}}]), \tag{13.13}$$

where the transition kernels $\mathbf{K}(t,x,[\pi_{\mathcal{V}}])$ is defined as:

$$\mathbf{K}(t,x,[\pi_{\mathcal{V}}]) = \mathbf{Q}(t,x) + \mathcal{I}(t,x,[\pi_{\mathcal{V}}]) + \mathbf{D}(t,x,[\pi_{\mathcal{V}}]). \tag{13.14}$$

Equation (13.13) can be solved applying *upwind semi-discretization* [22]. The partial differential equation is first discretized over the space, and then solved with respect to the time. The *upwind* technique makes a first-order approximation of the spatial derivative with respect to the direction of movements. As a result, we obtain a system of ordinary differential equations that can be solved using standard methods like Euler or Runge–Kutta.

AMMs can be exploited to study the behaviour of Intelligent Transportation Systems (ITS) [26], where information and communication technologies are applied to the design, analysis, monitoring and control of transportation systems, with particular emphasis on road networks. Transportation systems can involve multiple entities, such as humans, vehicles and the physical infrastructure, interacting each others. Several aspects of transportation systems are uncertain and nonlinear. They are usually large scale and are always geographically distributed. For all such reasons, AMMs are suitable to describe and study performance of the ITS in terms of road traffic congestion, safety and so on. An AMM was first applied to ITS for the quantitative risk analysis of collision of vehicles in a road tunnel [7]. The model describes the behaviour of a flow of vehicles capable to sense their proximity to other cars or to unexpected obstacles. According to such informations, the vehicles automatically adapt their speed in order to preserve a prescribed safety distance, or even brake to avoid collisions. In the model, MAs code in their space-state the cruise speed of vehicles. Proximity messages are periodically exchanged among MAs and influence such speeds as described by the $\mathcal{I}(t,x,[\pi_{\mathcal{V}}])$ term. The perception function rules the reaction of the vehicle to the proximity with other cars or obstacles. Birth term $\mathbf{b}(t,x,[\pi_{\mathcal{V}}])$ and death term $\mathbf{D}(t,x,[\pi_{\mathcal{V}}])$ account for the entrance and exit of vehicle in the tunnel, respectively. The results provided by the model were used to compute the values of the safety distance and the maximum allowed speed that minimize the probability of collisions inside the tunnel.

13.3.5 Population Models

Population Markovian Agents (PMAs) are Markovian Agents models where agents can move to other locations, can increase in number or decrease (either spontaneously or induced by other agents), or they can multiply during the transitions.

To simplify the presentation, we will consider a single class of agents, and we will drop subscript c or superscript $\{c\}$ from the equations. We will also focus on a discrete space model $\mathcal{V} = \{\mathbf{v}_1, \mathbf{v}_2, \ldots \mathbf{v}_N\}$. A PMA describes the evolution of a single agent, and it is defined by a tuple $(\mathcal{Q}, b, d, R, \eta, N)$. The first three components define the common Markovian Agent behaviour, while the last three account for the population evolution of the system. $\mathcal{Q} = (\widetilde{Q}(\mathbf{v}), \widehat{Q}^{[k]}(\mathbf{v}, \mathbf{v}'))$, are the *transition rate matrices*. $\tilde{q}_{ij}(\mathbf{v})$ is the rate at which an agent in location $\mathbf{v}$ jumps from state i to j due to local activities. $\widehat{Q}_{ij}^{[k]}(\mathbf{v}', \mathbf{v})$ accounts for transitions caused by inductions, and represents the rate at which an agent in state k in position $\mathbf{v}'$ induces jumps from state i to state j in position $\mathbf{v}$. In Eq. (13.4) we have

$$\mathbf{Q}_c(t, \mathbf{v}) = \widetilde{Q}(\mathbf{v}), \quad \mathcal{I}_c(t, \mathbf{v}, [\boldsymbol{\pi}_{\mathcal{V}}]) = \sum_{\mathbf{v}' \in \mathcal{V}} \sum_k \widehat{Q}^{[k]}(\mathbf{v}', \mathbf{v}) \boldsymbol{\pi}_k(t, \mathbf{v}'). \tag{13.15}$$

Component $b = (\tilde{b}(l), \hat{b}^{[k]}(\mathbf{v}, \mathbf{v}'))$, where $\tilde{b}(\mathbf{v}) = |\tilde{b}_i(\mathbf{v})|$ and $\hat{b}^{[k]}(\mathbf{v}) = |\hat{b}_i^{[k]}(\mathbf{v}', \mathbf{v})|$, are respectively the spontaneous and induced *births vectors*. In particular, $\tilde{b}_i(\mathbf{v})$ represents the rate at which agents are created in state i at location $\mathbf{v}$, and $\hat{b}_i^{[k]}(\mathbf{v}', \mathbf{v})$ is the rate at which an agent in state k in position $\mathbf{v}'$ induces birth of agents in state i at location $\mathbf{v}$. In a similar way, $d = (\tilde{d}(\mathbf{v}), \hat{d}^{[k]}(\mathbf{v}, \mathbf{v}'))$, where $\tilde{d}(\mathbf{v}) = |\tilde{d}_i(\mathbf{v})|$ and $\hat{d}^{[k]}(\mathbf{v}', \mathbf{v}) = |\hat{d}_i^{[k]}(\mathbf{v}', \mathbf{v})|$ are respectively the spontaneous and induced *death vectors*, where $\tilde{d}_i(\mathbf{v})$ represents the rate at which agents are destroyed in state i at location $\mathbf{v}$, and $\hat{d}_i^{[k]}(\mathbf{v}', \mathbf{v})$ is the rate at which an agent in state k and position $\mathbf{v}'$ induces decrease in the number of agents in state i at location $\mathbf{v}$. In Eqs. (13.3) and (13.4), we have that the birth and death kernels can be defined as

$$\mathbf{b}_c(t, \mathbf{v}, [\boldsymbol{\pi}_{\mathcal{V}}]) = \tilde{b}(\mathbf{v}) + \sum_{\mathbf{v}' \in \mathcal{V}} \sum_k \hat{b}^{[k]}(\mathbf{v}', \mathbf{v}) \boldsymbol{\pi}_k(t, \mathbf{v}'), \tag{13.16}$$

$$\mathbf{D}_c(t, \mathbf{v}, [\boldsymbol{\pi}_{\mathcal{V}}]) = \mathrm{diag}\left(\tilde{d}(\mathbf{v}) + \sum_{\mathbf{v}' \in \mathcal{V}} \sum_k \hat{d}^{[k]}(\mathbf{v}', \mathbf{v}) \boldsymbol{\pi}_k(t, \mathbf{v}') \right). \tag{13.17}$$

The peculiarity of PMAs are the set of *reactions* $R = \{r_1, \ldots, r_{|R|}\}$ that allows agents to move, duplicate or merge, either in the same or in neighbour locations. Reactions are characterized by a *reaction vector* $\eta(\mathbf{v}, [\boldsymbol{\pi}_{\mathcal{V}}]) = |\eta_h(\mathbf{v}, [\boldsymbol{\pi}_{\mathcal{V}}])|$, with $\mathbf{v} \in \mathcal{V}$. In particular, $\eta_h(\mathbf{v}, [\boldsymbol{\pi}_{\mathcal{V}}])$ represents the speed at which reaction r_h occurs in location $\mathbf{v}$. Since reaction rates can have very complex expressions, they can be the functions of the complete state of the model. The effects of reactions are described by $\mathcal{N} = (\mathcal{N}^+(\mathbf{v}, \mathbf{v}''), \mathcal{N}^-(\mathbf{v}, \mathbf{v}''))$ where $\mathcal{N}^+(\mathbf{v}, \mathbf{v}'') = |n_{i,k}^+(\mathbf{v}, \mathbf{v}'')|$ and $\mathcal{N}^-(\mathbf{v}, \mathbf{v}'') = |n_{i,k}^-(\mathbf{v}, \mathbf{v}'')|$ are $n_c \times |R|$ matrices that describe the changes on the number of agents due to the occurrence of the reaction $r_k \in R$. The value of $n_{i,k}^+(\mathbf{v}, \mathbf{v}'')$ represents the number of agents that are added to state i in location $\mathbf{v}''$ when reaction r_k takes place, while $n_{i,k}^-(\mathbf{v}, \mathbf{v}'')$ accounts for the agents that are removed. The formalization of terms η and $\mathcal{N}$ is quite similar to one used in computational system biology (see [14]). Reactions can also multiply or divide the number of agents in a given location. In this case, both $n_{j,e}^-(\mathbf{v}, \mathbf{v}'') > 0$ and

$n^+_{j,e}(\mathbf{v}, \mathbf{v}'') > 0$ for the same reaction r_e and state j. The definitions of η and $\mathcal{N}$ are expressive enough to model several different features: duplication or decimation of agents in a state, movement of agents (deterministic or probabilistic) to other locations, duplication or decimation combined with state jump and location movement. Interested readers can find more details in [11]. Reactions can be implemented in Eq. (13.1) by defining the input and output kernels as follows:

$$\mathbf{m}_c^{in}(t, \mathbf{v}, [\boldsymbol{\pi}_\mathcal{V}]) = \sum_{\mathbf{v}'' \in L} \eta(\mathbf{v}'', [\boldsymbol{\pi}_\mathcal{V}])(N^+(\mathbf{v}'', \mathbf{v}))^T, \tag{13.18}$$

$$\mathbf{M}_c^{out}(t, \mathbf{v}, [\boldsymbol{\pi}_\mathcal{V}]) = \mathrm{diag}(\boldsymbol{\pi}(t, \mathbf{v}))^{-1} \sum_{\mathbf{v}'' \in L} \eta(\mathbf{v}, [\boldsymbol{\pi}_\mathcal{V}])(N^-(\mathbf{v}, \mathbf{v}''))^T, \tag{13.19}$$

where $(\mathcal{N})^T$ denotes the transpose of matrix $\mathcal{N}$. Equation (13.18) considers agents that are entering a location $\mathbf{v}$, coming from a reaction that takes place in a location $\mathbf{v}''$. Equation (13.19) accounts for the change in the number of agents occurring in one state due to elements removed by the reactions. In particular, it accounts for all the agents that are leaving location $\mathbf{v}$ directed to a reaction happening in location $\mathbf{v}''$. Since the rate at which reactions occurs already accounts for the density of the agents, in order to allow the inclusion in the output kernel of Eq. (13.4), we must normalize the effect by multiplying on the left by $\mathrm{diag}(\boldsymbol{\pi}(t, \mathbf{v}))^{-1}$.

The definition of $\eta(\mathbf{v}, [\boldsymbol{\pi}_\mathcal{V}])$ must depend on the total state of the system $[\boldsymbol{\pi}_\mathcal{V}]$, and it must be correctly defined to prevent reactions to happen when there are not enough agents in the involved states: an improperly defined function $\eta(\mathbf{v}, [\boldsymbol{\pi}_\mathcal{V}])$ can lead to negative counts in the number of agents. Determining the condition for which a function $\eta(\mathbf{v}, [\boldsymbol{\pi}_\mathcal{V}])$ does not lead to negative counts, is an important topic that will be investigated in future woks: here we will limit ourselves to consider functions coming from system biology, that are known to correctly behave as reaction rates. In particular we refer to the *Law of Mass Action* which was studied by Waage and Guldberg in 1864. Such law tells that the reaction rate is proportional to the probability of a collision of the reactants, that in turn is proportional to the concentration of reactants (agents in our case), elevated to the multiplicity required to start the reaction:

$$\eta_h(\mathbf{v}, [\boldsymbol{\pi}_\mathcal{V}]) = \gamma_h \prod_{j:\mathcal{V}^-(\mathbf{v},j) \neq \emptyset} \left(\sum_{\mathbf{v}'' \in \mathcal{V}^-(\mathbf{v},j)} \pi_j(t, \mathbf{v}') \right)^{\sum_{\mathbf{v}'' \in \mathcal{V}^-(\mathbf{v},j)} (n^-_{j,h}\mathbf{v}, \mathbf{v}'')}. \tag{13.20}$$

Here γ_h represents the speed at which the reaction occurs. Note that since the *source* agents of reaction r_h in location $\mathbf{v}$ can arrive from any location $\mathbf{v}''$ such that $n^-_{j,h}(\mathbf{v}, \mathbf{v}'') > 0$, (here denoted with $\mathcal{V}^-(\mathbf{v}, j) = \{\mathbf{v}' \in \mathcal{V} : n^-_{j,h}(\mathbf{v}, \mathbf{v}') > 0\}$), the total count of agents required to engage a reaction must be computed with the sum $\sum_{\mathbf{v}'' \in \mathcal{V}^-(\mathbf{v},j)} n^-_{j,h}(\mathbf{v}, \mathbf{v}'')$. For the same reason, the number of agents involved in the reaction, must be computed with the sum over all the possible input locations, i.e. $\sum_{\mathbf{v}' \in \mathcal{V}^-(\mathbf{v},j)} \pi_j(t, \mathbf{v}')$.

PMAs are applied in [11] to analyze the cancer evolution, and to comprehend the mechanisms underlying the Cancer Stem Cells hierarchy whose characterization is crucial in the study of tumour progression. Current evidence indicates that many cancers arise from a small population of cells named Cancer Stem Cells (CSCs), which have undergone malignant transformation driven by frequent genetic mutations. The application of the PMAs to describe the behaviour of such cell populations allowed the scientists to observe their proliferation through the host tissue. The model is exploited to consider movement of cell populations in a bi-dimensional space, and it is used to derive the system evolution.

Recent studies in cancer biology have led to a new perspective in tumour progression, known as the CSC theory. It states that the growth and evolution of many cancers are driven by a small population of cells named CSC, and that CSC-based tumours are hierarchically structured, and characterized by different subpopulations of cells: CSCs, Progenitor Cells (PCs) and Totally differentiated Cells (TCs). Moreover, such heterogeneity is considered the cause of the failure of many conventional therapies. It is hence fundamental to fully comprehend the CSC CSC dynamics to predict treatment response. The proposed PMA model describes the CSC-based tumour growth and it is able to reproduce the overall dynamics among cell subpopulations during tumour progression. Using a derivation similar to the Generalized Mass Action law, the reactions describing the biological model were translated into a system of Ordinary Differential Equations (ODEs) able to take into account the possibility of reactions to expand in neighbour cells.

13.3.6 Abstract Space Models

When space $\mathcal{V} = \{\mathbf{v}_1, \mathbf{v}_2, \ldots\}$ is discrete, it can be used not only to describe graph-based routes, but also to represent more abstract configurations. For example, in [5] locations are used to model different data centres of a geographically distributed cloud infrastructure. In that case $\mathcal{V} = \{dc_1, dc_2, \ldots\}$ locations are used to model the different data centres composing the infrastructure. Dependency on the location is used to assign different capabilities in terms of computational nodes that are able to run Virtual Machines (VMs), and disk infrastructure capable of saving data as Storage Blocks (SBs).

Agents classes $1 \leq c \leq C$ represents applications running in the data centre: the state of the agent defines the resource usage of each application in the infrastructure. The number of applications for each class c running at data centre dc_j is encoded into the agent density function $\rho_c(t, dc_j)$. The average performance of the nodes of the data centre is encoded in the transition function $\widetilde{\mathbf{K}}_c([\boldsymbol{\pi}_\mathcal{V}])$ for each application class. In this case, in particular, the local transition kernel $\mathbf{Q}_c(t, \mathbf{v}) = 0$ since the speed at which application acquires and releases resources depends on the entire state of the data centre, and $\mathcal{I}_c(t, \mathbf{v}) = \widetilde{\mathbf{K}}_c([\boldsymbol{\pi}_{\mathcal{V}}])$.

If we consider a fixed number of applications, the birth term and death term can be set to $\mathbf{b}_c(t, \mathbf{v}, [\boldsymbol{\pi}_{\mathcal{V}}]) = 0$ and $\mathbf{D}_c(t, \mathbf{v}, [\boldsymbol{\pi}_{\mathcal{V}}]) = 0$, otherwise they can be used to model the starting and stopping of applications in presence of fluctuations in the workload. The motion terms $\mathbf{M}_c^{out}(t, \mathbf{v}, [\boldsymbol{\pi}_{\mathcal{V}}])$ and $\mathbf{m}_c^{in}(t, \mathbf{v}, [\boldsymbol{\pi}_{\mathcal{V}}])$ can be used to model application migration from one data centre to the other when supporting load-balancing applications that works at the complete geographical infrastructure level.

Abstract space models have also been used in [19] to model cell hand-over and technology switch in wireless network models, where user agents can connect to either a 4G LTE network or to a WiFi access point depending on the current load.

13.3.7 Delay Models

The ability of agents that perceive the state of the model even at previous time instants $[\boldsymbol{\pi}_{\mathcal{V}}]$, can be used to add random or deterministic delays in the delivery of messages to the model presented in Sect. 13.3.2. In particular, let us assume that the delivery of message μ sent from a class c agent in position $\mathbf{v}$ to a class c' agent in position $\mathbf{v}'$ during a transition from state i to state j required a random time τ distributed according to a positive continuous distribution $Y_\mu(\tau|\mathbf{v}, c, i, \mathbf{v}', c', j)$. The time delay in messages can be considered by replacing the definition of $\gamma_{ii}^{\{c\}}(t, \mathbf{v}, [\boldsymbol{\pi}_{\mathcal{V}}], \mu)$ previously given in Eq. (13.8) with:

$$\gamma_{ii}^{\{c\}}(t, \mathbf{v}, [\boldsymbol{\pi}_{\mathcal{V}}], \mu) = \int_{\substack{\mathbf{v}' \in \mathcal{V} \\ \mathbf{v}' \neq \mathbf{v}}} \sum_{c'=1}^{C} \sum_{j=1}^{n_{c'}} \beta_j^{\{c'\}}(\mathbf{v}', \mu) u_\mu(\mathbf{v}, c, i, \mathbf{v}', c', j) \int_0^t \pi_j^{\{c'\}}(t - \tau, \mathbf{v}') dY_\mu(\tau|\mathbf{v}, c, i, \mathbf{v}', c', j) d\mathbf{v}'. \tag{13.21}$$

In this new definition, the time integral over τ accounts for the state of the system at time $t - \tau$, weighted by the probability that the delay distribution $Y_\mu(\tau|v, c, i, \mathbf{v}', c', j)$ is equal to τ. This definition makes Eq. (13.1), which normally is either an ordinary differential equation (if motion over a continuous space is not considered) or a partial differential equation, a *delay differential equation*. Although the techniques for the solution of delay differential equation are very solid, they can be computationally expensive and reduce the size of models that could be fruitfully analyzed. From a theoretical point of view, however, this result enhances the possibility of MA modelling: a full investigation of the advantages of using delay differential equation models, and the definition of the use cases when this can provide good results is ongoing research work.

13.4 Conclusions

In this work, we have presented a unified theory that can be used to study various types of Markovian Agent models. In particular, we have shown that a single equation can describe all the peculiarities of agents: local transitions, induced behaviours, motion, increase and decrease in population. Markovian Agents have been used to study a variety of different systems, both physical and IT related: in this work we have provided some hints on how the formalism has been used, and given links to the literature where such cases have been discussed.

Even if the theory is already solid, many features and limitations of Markovian Agents still need to be investigated: future works will focus on improving solution techniques, providing tools to simplify the use of MA models, defining best-practices and study the scalability and accuracy of MA models.

References

1. Ball F, Milne R, Tame I, Yeo G (1997) Superposition of interacting aggregated continuous-time Markov chains. Adv Appl Probab 29:56–91
2. Bobbio A, Gribaudo M, Telek M (2008) Analysis of large scale interacting systems by mean field method. In: 5th international conference on quantitative evaluation of systems—QEST2008. St. Malo
3. Bruneo D, Scarpa M, Bobbio A, Cerotti D, Gribaudo M (2012) Markovian agent modeling swarm intelligence algorithms in wireless sensor networks. Perform Eval 69:135–149
4. Bruneo D, Scarpa M, Bobbio A, Cerotti D, Gribaudo M (2015) An intelligent swarm of markovian agents. In: Springer handbook of computational intelligence, pp 1345–1359
5. Castiglione A, Gribaudo M, Iacono M, Palmieri F (2015) Modeling performances of concurrent big data applications. Softw Pract Exp 45(8):1127–1144. doi:10.1002/spe.2269
6. Cerotti D, Gribaudo M, Bobbio A (2009) Disaster propagation in inhomogeneous media via markovian agents. In: Critical information infrastructure security, vol 5508, pp 328–335. Springer LNCS (2009)
7. Cerotti D, Gribaudo M, Bobbio A (2009) Presenting dynamic markovian agents with a road tunnel application. In: 17th annual meeting of the IEEE/ACM international symposium on modelling, analysis and simulation of computer and telecommunication systems, MASCOTS 2009, September 21–23, 2009, South Kensington Campus, Imperial College London, pp 1–4
8. Cerotti D, Gribaudo M, Bobbio A (2014) Markovian agents models for wireless sensor networks deployed in environmental protection. Rel Eng Syst Saf 130:149–158
9. Cerotti D, Gribaudo M, Bobbio A, Calafate CMT, Manzoni P (2010) A markovian agent model for fire propagation in outdoor environments. In: Computer Performance Engineering—7th European Performance Engineering Workshop, EPEW 2010, Bertinoro, Italy, September 23–24, 2010. Proceedings, pp 131–146
10. Ciardo G, Muppala J, Trivedi K (1991) On the solution of GSPN reward models. Perform Eval 12:237–253
11. Cordero F, Fornari C, Gribaudo M, Manini D (2014) Markovian agents population models to study cancer evolution. In: Analytical and stochastic modelling techniques and applications—21st international conference, ASMTA 2014, Budapest, Hungary, June 30–July 2, 2014. Proceedings, pp 16–32

12. Deneubourg Aron S, Goss S, Pasteels JM (1990) The self-organizing exploratory pattern of the argentine ant. J Insect Behav 3(2):159–168. doi:10.1007/BF01417909
13. Dorigo M, Stützle T (2004) Ant colony optimization. MIT Press
14. Edwards J, Palsson B (1998) How will bioinformatics influence metabolic engineering? Biotechnol Bioeng 58(2–3):162–169
15. Gilmore S, Hillston J, Kloul L, Ribaudo M (2003) PEPA nets: a structured performance modelling formalism. Perform Eval 54(2):79–104. http://www.sciencedirect.com/science/article/pii/S0166531603000695
16. Goss S, Aron S, Deneubourg J, Pasteels J (1989) Self-organized shortcuts in the Argentine ant. Naturwissenschaften 76(12):579–581. doi:10.1007/BF00462870
17. Gribaudo M, Cerotti D, Bobbio A (2008) Analysis of on-off policies in sensor networks using interacting Markovian agents. In: 4th international work sensor networks and systems for pervasive computing—PerSens 2008, pp 300–305
18. Gribaudo M, Chiasserini CF, Gaeta R, Garetto M, Manini D, Sereno M (2005) A spatial fluid-based framework to analyze large-scale wireless sensor networks. In: IEEE international conference on dependable systems and networks, DSN2002
19. Gribaudo M, Manini D, Chiasserini C (2013) Studying mobile internet technologies with agent based mean-field models. In: Analytical and stochastic modelling techniques and applications—20th international conference, ASMTA 2013, Ghent, Belgium, July 8–10, 2013. Proceedings, pp 112–126
20. Guenther, M.C., Bradley, J.T.: Higher moment analysis of a spatial stochastic process algebra. In: Computer Performance Engineering, pp. 87–101. Springer - LNCS, Vol 6977, Springer (2011)
21. Hillston J (2005) Fluid flow approximation of PEPA models. In: 2nd international conference on quantitative evaluation of systems—QEST, pp 33–43
22. Horton G, Kulkarni VG, Nicol DM, Trivedi KS (1998) Fluid stochastic petri nets: theory, application, and solution techniques. Eur J Oper Res 105(1):184–201
23. Kaâniche M, Lollini P, Bondavalli A, Kanoun K (2012) Modeling the resilience of large and evolving systems. CoRR abs/1211.5738
24. Kurtz T (1981) Approximation of population processes. Soc Ind Appl Math. http://epubs.siam.org/doi/abs/10.1137/1.9781611970333
25. Plateau B, Atif K (1991) Stochastic automata network for modeling parallel systems. IEEE Trans Softw Eng 17:1093–1108
26. Sheng-hai A, Byung-Hyug L, Dong-Ryeol S (2011) A survey of intelligent transportation systems. In: 2011 Third international conference on computational intelligence, communication systems and networks (CICSyN), pp 332–337
27. Sukhwani H, Bobbio A, Trivedi K (2015) Largeness avoidance in availability modeling using hierarchical and fixed-point iterative techniques. Int J Perform Eng 11(4):305–319
28. Tomek L, Trivedi K (1991) Fixed point iteration in availability modeling. In: Cin M, Hohl W (eds) Fault-Tolerant Computing Systems, *Informatik-Fachberichte*, vol 283. Springer, Berlin, pp 229–240
29. Trivedi KS (2002) Probability and statistics with reliability, queuing and computer science applications. Wiley, Chichester
30. Trivedi KS, Ma X, Dharmaraja S (2003) Performability modelling of wireless communication systems. Int J Commun Syst 16(6):561–577. doi:10.1002/dac.605

Index

K. Al-Begain and A. Bargiela (eds.), *Seminal Contributions to Modelling and Simulation*, Simulation Foundations, Methods and Applications,
DOI 10.1007/978-3-319-33786-9

Printed in the United States
By Bookmasters